Lecture Notes in Morphogenesis

Lecture Notes in Morphogenesis is an interdisciplinary book series which aims to face the questions of emergence, individuation and becoming of forms from several different points of view: those of pure and applied mathematics, of computational algorithms, of biology, of neurophysiology, of cognitive and social structures. The set of questions above concerns all the manifestations of Being, all the manifestations of Life. At the heart of contemporary embryogenesis lies an essential question: How can form emerge from the constant, chaotic flow? How can a sequence of purely informational elements — an a-signifying combination of chemical substances organized in the DNA molecule — evolve into the highly complex and structured forms of the living organism? A similar question can be asked when we deal with the morphogenesis of vision in neural systems and with the creation of evolving synthetic images, since digital technology makes possible the simulation of emergent processes both of living bodies and of visual forms. Finally the very idea that abstract structures of meaning could be captured in terms of morphodynamic evolution opens the door to new models of semiolinguistics, semiotic morphodynamics, and cognitive grammars. An entire heritage of ideas and concepts has to be reconsidered in order to face new and challenging problems: the theoretical framework opened by Goethe with the introduction of the word "Morphogenesis" is developed by D'Arcy Thompson in "On Growth and Form", it is reorganized with new theoretical insights by the classical structuralism of Levi-Strauss and formalized by the dynamical structuralism of René Thom. The introduction of the post-structuralists ideas of individuation (in Gilbert Simondon and Gilles Deleuze) and plasticity of structures builds a bridge to contemporary problems of morphogenesis at a physical, biological, social and transindividual level. The objective of this book series is to provide suitable theoretical and practical tools for describing evolutionary phenomena at the level of Free boundary problems in Mathematics, Embryogenesis, Image Evolution in Visual Perception, Visual Models of Morphogenesis, Neuromathematics, Autonomy and Self-Organization, Morphogenetic Emergence and Individuation, Theoretical Biology, Cognitive Morphodynamics, Cities Evolution, Semiotics, Subjectivation processes, Social movements as well as new frontiers of Aesthetics. To submit a proposal or request further information, please use the PDF Proposal Form or contact directly: *Dr. Thomas Ditzinger (thomas.ditzinger@springer.com)*

Wolfgang Wildgen

Morphogenesis of Symbolic Forms: Meaning in Music, Art, Religion, and Language

Springer

Wolfgang Wildgen
Institut für Allgemeine und Angewandte
Sprachwissenschaft
Universität Bremen
Bremen, Germany

ISSN 2195-1934 ISSN 2195-1942 (electronic)
Lecture Notes in Morphogenesis
ISBN 978-3-031-25650-9 ISBN 978-3-031-25651-6 (eBook)
https://doi.org/10.1007/978-3-031-25651-6

This Springer imprint is published by the registered company Springer Nature Switzerland AG
The registered company address is: Gewerbestrasse 11, 6330 Cham, Switzerland

In memory of René Thom (1923–2002) to the centenary of his birth
To my grandson Ludwig Vito Wildgen, born on the 29th of September 2021

Preface

The central aim of this research is to give a morphogenetic, dynamic view of human communication, culture, and symbolic media. The underlying theoretical innovation originated in the work of René Thom (1923–2002) and Hermann Haken (1927–). The philosophical background refers to the *Philosophy of Symbolic Forms* by Ernst Cassirer (1874–1945). His focus was on language, myth, and science.

In the present book, the starting line is defined by a morphogenetic perspective, mainly on human communication and culture's evolution and further development. A second and significant focus is on visual communication, music, and religion (myth), i.e., on the "symbolic forms" (Cassirer) beyond language. The term "morphogenesis" has more precisely the meaning given by René Thom (1923–2002) in his publications on "Morphogenesis and Structural Stability" (1972) and "Semiophysics" (1988). Finally, the applications of Synergetics in the Humanities (cf. Haken and Stadler 1989, Kelso 1995, and Haken 1996) and the notions of "self-organization" and cooperation or rivalry between subsystems enlarge the scope of our analysis.

Visual art from the Paleolithic to modernity may stand for visual communication. The present book focuses on studies of classical painting and sculpture (e.g., Leonardo da Vinci, William Turner, and Henry Moore) and modern art (e.g., the sculptures of William Moore and Joseph Beuys and the paintings of Jackson Pollock). Musical morphogenesis embraces classical music (from J. S. Bach to Arnold Schönberg), songs, and political songwriting. In the case of religion, the myths of pre-literary and pre-industrial societies show essential gradients of self-organization, mainly in the re-assembly of traditions. Classical polytheistic and monotheistic religions also demonstrate language and logic's effect as guiding symbolic forms. Finally, significant trends can be analyzed in the case of Christianism.

The book selects and elaborates topics dealt by the author in his trilogy published in German: (titles of these books translated into English) 2013, "Visual Semiotics. The Unfolding of the Visible. From Cave Art to Modern Towns", 1918, "Musical Semiotics. Musical Signs, Cognition, and Language"; 2021 "Myth and Religion. Semiotics of the Transcendent".

Some representative analyses are developed further and exemplify the use and the scientific relevance of the morphogenetic paradigm issued by René Thom and Hermann Haken. A new cultural and symbolic philosophy concept emerges going beyond Peirce and Cassirer and based on principles developed in the natural sciences.

Bremen, Germany Wolfgang Wildgen

Acknowledgments Earlier versions of chapters in this book have been commented on by Prof. Dr. Martina Plümacher (Humbolt University, Berlin) and Dr. Ian Verstegen (University of Pennsylvania). In addition, corrections to English style and language have been contributed by Prof. Dr. Clausdirk Pollner (University of Leipzig), Dr. Janina Wildfeuer (Rijksuniversiteit Groningen), and Dr. Ian Verstegen (Chap. 4).

Thanks to Heidemarie Wildgen for her support during the manuscript preparation, which absorbed most of my free time in 2022.

References

Haken, H., Stadler, M.: Synergetics of Cognition. Proceedings of the International Symposium at Schloss Elmau, Bavaria, June 4–8, 1989. Springer, Berlin (1990)

Haken, H.: Principles of Brain Functioning. A Synergetic Approach to Brain Activity, Behavior, and Cognition. Springer, Berlin (1996)

Kelso, J. A. S.: Dynamic Patterns. The Self-organization of Brain and Behavior MIT Press, Cambridge (Mass.) (1995)

Thom, R.: Stabilité structurelle et morphogenèse, Interéditions, Paris (English translation: Structural stability and morphogenesis. Benjamin, Reading, 1975) (1972) Thom R. : Esquisse d'une sémiophysique: physique aristotélienne et théorie des catastrophes. Interéditions, Paris [English translation: Semiophysics. A Sketch. Addison-Wesley, Boston 1990] (1988)

Contents

Chapter 1
Introduction

Abstract Morphogenesis, the general topic of the Lecture Notes, has been applied to language and other cultural media (symbolic forms). An essential input came from René Thom, the French mathematician and Fields Medal winner. The chapter starts by considering the role of morphogenesis in the life sciences and the transition from biological to semiotic morphogenesis (semiogenesis). Further elaborations consider the impact of individuation, the specification of morphogenesis for human populations and individuals, the role of traditions and reflection (ratiogenesis), and, eventually, the relation between morphogenesis (in the sense of Thom 1972) and "semiophysics" (exposed in (Thom in Esquisse d'une sémiophysique: physique aristotélienne et théorie des catastrophes. Interéditions, Paris [English translation: Semiophysics. A Sketch. Addison-Wesley, Boston 1990], 1988 [20]). Parallel to Thom's proposals, the interdisciplinary field of Synergetics, introduced by Herman Haken, has widened the consideration by the consideration of stochastic dynamics, the analysis of cooperative effects between systems, and the complexities of self-organization in nature and culture.

1.1 Morphogenesis and the Science of Life

The most basic biology question concerns the transition from inert matter to life. In this transition, physical laws do not lose their relevance; on the contrary, they are necessary to explain the origin of life. Nevertheless, essential features change dramatically. Physical laws are reversible, and phase transitions are punctual. In contrast, the transitions and bifurcations in the domain of life are irreversible (until death), processes in the organism depend on the ecology of the system, negative entropy (information gain) is possible,[1] and different levels of organization (with specific laws) are eminent. Minor causes and chance effects may decide upon the further

[1] See the concept of anti-entropy that includes not only a gain of order but also its maintenance in Longo and Montévil ([11]: 19f and 254).

W. Wildgen, *Morphogenesis of Symbolic Forms: Meaning in Music, Art, Religion, and Language*, Lecture Notes in Morphogenesis,
https://doi.org/10.1007/978-3-031-25651-6_1

history of a group or even a species (Blount [3] calls this effect “contingency”), convergent evolutions lead to similarities between species, which are not the result of an identical genetic outfit or parallel causal lines. With the advent of theoretical biology, mathematical technics were inspired by their applications in physics, although the underlying mechanisms are different. These differences ask for a modification of the formal tools applied in physics. Although, in the beginning, biological systems (unicellular organisms) may still be accessible to a physicochemical analysis, complex biological systems, specifically the building of life-forms and their differentiation, ask for an independent treatment that respects the differences mentioned above and considers holistic phenomena, e.g., the telic character of wholes and the back-propagation from the whole to its parts. This aspect had already been clear to Kant at the end of the eighteenth century (“organisms are wholes”). Philosophers of the nineteenth century (e.g., Hegel) made clear that the aspect of genesis, of development, is at the heart of biological reasoning. The morphogenetic phenomena are, therefore, crucial for biology. However, are they also vital to psychology, sociology, and semiotics?

In psychology and the cognitive sciences, we must consider a new transition, which at first sight seems to be inside biology. Living beings interact with the ecology and have their inner stability (metabolism). The brain, characteristic of higher animals, is part of the whole organism. Nevertheless, the brain’s functions in perception, communication, and interaction open a field of processes beyond purely biological morphogenesis. Are phenomena like thinking, (self)-consciousness, and symbolic behavior still accessible via the mechanisms found in the organization of organic life? Because of continuity in evolution, we should assume that on one side basic types of organization in cognitive, communicative, and semiotic systems are shared with the organization of primary biotic systems; on the other side, new features can emerge, establishing a kind of autonomy of symbolic systems. In the context of our topic, we must ask: Can we still describe morphogenesis and the stability and regulation of forms of behavior in the framework of theoretical biology? It seems plausible to conjecture a transition similar to the one considered when the physical frame changed into a biological one. As in the former case, some general features will remain relevant, but new phenomena, regularities, and laws should appear.

1.1.1 The Rise of Theoretical Biology and the Role of Morphogenesis

The theoretical treatment of morphogenesis, specifically on mathematical background, goes back to D’Arcy Thompson (1860–1948), who in 1917 published his book *On Growth and Form*. Significant advances were due to Conrad Waddington (1905–1975) and his work in the thirties on “developmental epigenetics" (cf. his book: Organizers & Genes, 1940). René Thom (1923–2002) carried this thread on. He used mathematical results of differential topology (catastrophe theory) to map the processes of biological morphogenesis and expanded this view to the human and social sciences (cf. Thom [19]). In the following decades, this program was further substantiated and diversified, mainly using the quickly evolving field of dynamic systems theory (catastrophes, chaos, fractals, stochastic dynamics). Most of the proponents of this endeavor came from mathematics (Thom and Zeeman), physics (Haken), and chemistry (Prigogine). In the eighties, the concepts of autopoiesis and self-organization advocated by the neuroscientist Humberto Maturana (*1928) attracted many biologists and psychologists (cf. Maturana and Varela [12]).

Meanwhile, significant results and theoretical advances have been integrated into the field of the corresponding natural sciences (from physics to the neurosciences). However, applications to the human and social sciences are still controversial and ask for a better empirical and theoretical foundation. If Johann Wolfgang Goethe (1749–1832) considered an embracing discipline "Morphologie überhaupt" (universal morphology) ranging from biology to the arts (literature, visual art, music), this dream seems still to be far off. Intuitively, the creation of forms, their classification, their historical evolution, and the artist's creative act ask for concepts similar to those found in the analysis of living beings. Beyond intuitive creativity, art concerns primarily the emotion, the will, and the rationality of a human being in the creation and appraisal of symbolic forms.

1.1.2 Darwin Revisited

The first effect of Darwinism, responsible for its triumph after almost a century of debates, was that its theory rendered God as an explanatory force of evolution superfluous. As such, it continued the trend established by Condillac, Rousseau, and Herder in the eighteenth century and announced the triumph of physical science at the end of the nineteenth century. Its second impulse was to introduce a space of possible evolutions (variation in Darwin's terms) out of which selection can choose. This space became more concrete with modern genetics, the explanation of mutations, and the deciphering of DNA. In addition, statistical models could model the interaction of mutation and selection. These advances transformed Darwin's incomplete explanation into Neo-Darwinism (the New Synthesis).

A further transformation of Darwinism occurred in the late seventies of the twentieth century and led to new theories called EVO-DEVO (ecological evolutionary developmental biology). In this context, it was possible to revitalize fundamental insights of nineteenth-century biology, i.e., deep homologies that link biological entities separated by hundreds of millions of years, for instance, fruit flies, mice, and humans. The key to these analogies is ancient genes that have been conserved for millions of years and can control the homology of basic morphological "plans" despite huge morphological differences. Homeotic genes, i.e., genes that control the identity and correct order of bodily segments, constitute the "homeobox", i.e., the set of gene segments shared by the homeotic genes (cf. Gehring [7]: Chap. 3). Many of the homologies mentioned by René Thom in 1972 between patterns in plants and animals on the one side and humans and human behavior on the other can be related to such ancient genetic programs (morphogens, instructed signals).[2] Our discussion cannot go into the details of Evo–Devo-theories or experimental research in biochemistry. Still, this specification of Darwinian theory demonstrates that the morphological intuitions of René Thom and many predecessors (e.g., Goethe or St. Hilaire) are worth further consideration and do not contradict modern evolutionary biology.

In the case of symbolic forms, this means that although experimental research in genetics cannot bridge the gap between unity and diversity in fruit flies and universality/diversity in language and other symbolic forms, it makes sense to consider two different types of dynamics. First, the dynamics of self-organization start from a homogeneous base situation and arrive at a highly organized end situation. Mathematically, these dynamics refer to Turing's models in the early fifties of the twentieth century.[3] They may explain the rapid growth of diversity but cannot explain the long-lasting stability of species and families of species and their identity. Second, the existence of a developmental genetic toolkit that enables the conservation of old patterns in the midst of growing diversity (cf. Hidalgo et al. [9]). It remains still unclear what the nature of such a toolkit could be that controls the unity and diversity of cultural universals versus the divergence of cultural forms throughout the world. In Chap. 2, we shall argue that the organization of perceptual organs (in humans, particularly sight and hearing) constitutes a kind of "homeobox" for unfolding symbolic forms. A further ancient base is the dynamics of motion and locomotion, as the main Chaps. 3, 4, 5, and 6 will show.

Significant problems with evolutionary explanations remain. First, selection concerns the phenotype (the individual animal, its survival, and the creation of progeny); the transmission of the results to the genotype is less direct than presumed (epigenetic and global physiological factors interfere).[4] Second, the survival and expansion of new variants (based on unpredictable mutations) depend on many ecological and social contexts. Third, some effects may be self-referential as new variants change the survival conditions. This effect can occur *inside* the organism if variants trigger structural reorganizations or *outside* if moving and very active animals change their environment.[5] This effect became clear with humans and, most

---

[2] Mechanistic models of morphogenesis insist on the locality of all mechanisms and a reduction to physical and chemical action and reaction based on molecules. As our perspective is clearly "high-level", we do not adopt such a strictly molecular and local strategy. The Nobel prize winner Christiane Nüsslein-Vollhard shows in her book, Nüsslein-Vollhard (2004: 82–101), that basic processes of segmentation and spatial organization encountered in very different classes of animals establish gradients whose interaction and conflict are constitutive for morphogenesis.

[3] "Alan Turing first formulated this theory in 1952 with a 'reaction–diffusion' model describing the interaction of an activator and long-range diffusing inhibitor. Most work has since assumed a molecular basis for self-organization." Bailleul et al. ([2]: Sect. 1.3).

[4] Darwin still accepted a kind of Lamarckian mechanism which transforms individual cognitive and behavioral advances into features of a species.

[5] See also the notion of "ecomorph" introduced by Williams: a "species with the same structural habitat/niche, not necessarily close phyletically, but similar in morphology and behavior." (cited in Blount [3]).

dramatically, since the industrial revolution, which led to a new stage, the so-called *Anthropocene*, a world profoundly changed by man (Anthropos). These shortcomings of Darwinism have further consequences for evolutionary psychology, linguistics, and the evolution of human culture because complicated self-organization and self-referential effects occur in highly complex and large systems. Simple Darwinian mechanisms are more transparent and more decisive at the level of molecular evolution (cf. Eigen and Schuster [6]).

In the case of language evolution (cf. Wildgen 2004a), three relatively rapid and vital changes occurred:

(a) The transition from ape-like behavior in australopithecines (3–4 million years (=my) before present (=BP))[6] to Homo ergaster/erectus (ca. two my BP). It doubled cranial capacity, introduced tool use, and led to larger social groups.
(b) The transition from descendants of Homo ergaster in Africa to Homo sapiens finished ca. 300,000 years ago. After a bottleneck around 120,000 BP, this species expanded in and out of Africa (since 100,000, mainly after 70,000 BP). It seems plausible to explain these and further changes by applying morphogenetic principles.[7] In the case of language, this process led from the (hypothetical) protolanguage of Homo erectus to the human language capacity shared by all humans.
(c) Sophisticated cultural innovations appeared with the Neolithic revolution (beginning after the last ice age, i.e., approximately 10,000 BP), the first large and highly organized civilizations in Egypt and Mesopotamia, and the emergence of writing. However, it is not plausible that evolutionary processes affected essential human capacities in this period because the Darwinian mutation and selection mechanisms ask for more extensive time intervals due to mutation rates and the necessary time for the distribution and dominance of selected variants.

The communicative and social capacities of the evolving species (cf. Sect. 1.1.3) expand biological *morphogenesis* to *semiogenesis*. One must also consider *individuation*, which operates in the transition from genotype, the genetic type, to phenotype, the individual body/mind. Individuation contributes to the constitution of social systems like clans, regional associations of social groups, ethnical wholes, and large societies. These dynamics are responsible for the vast diversity of human cultures and symbolic forms. This process is called *ecological* and *social morphogenesis*; it concerns the unfolding of behavioral and cognitive "forms" (Greek *morphḗ*) in individuals and societies, their selection, and stabilization in an ecological context.

[6] BP = before present. "present" refers to a conference in 1955 where this category was created. In the case of historical time, the classical labels BC = before Christ, and AC = after Christ are used. Roughly, BC = BP – 2000.

[7] The last surviving non-human species was the late Neanderthal man in Europe (extinct between 37,000 and 30,000 BP), i.e., before the maximum of the last ice age (25,000 to 20,000 BP).

1.1.3 From Biological Morphogenesis to Semiogenesis

In epigenetic processes and embryology, a structural framing occurs, which limits new genetic variants and the bodily expression of genes, i.e., it shapes the space of possible forms. In the social, cultural, and ecological domain, other morphogenetic processes occur, called *semiogenetic*, insofar as the perception and mental reaction to the environment are a requisite of form-giving and form-transmission (imitation and learning). Cultural transmission replaces or goes parallel to genetic transfer. Semiogenesis redefines the relevant environment, changes the selective forces, and thus indirectly influences the genetic outfit and its epigenetic expression. Two highly relevant transitions after hominization had a major impact on the further development of human symbolic media and languages:

(a) The culture of painted caves in the late Paleolithic is documented between 37,000 and 16,000 BP. A rich corpus of paintings, drawings, sculptures, and abstract (quasi-writing) symbols illustrate this period, which extended over Central and Eastern Europe under the conditions of the last ice age (recently, cave paintings were discovered in Borneo).
(b) The new technologies of farming and cattle breeding led to many cultural innovations, e.g., writing and urbanization. It produced the first large-scale societies in Egypt, Mesopotamia, and the Indus valley (beginning around 5,000 BP = 3,000 BC).

A recent transition to a so-called "morphogenic society" in late modernity is advocated by Archer [1]. Earlier societies were relatively static, as negative feedback to social changes dominated and evinced rapid changes. In morphogenic societies, changes receive positive feedback. At first, innovations in social relations are realized in sub-cultures and social niches. After this experimental stage, some are rapidly generalized to the whole society or even globally. In a morphogenic society, symbolic media will also change rapidly and either lose their relevance and the number of adherents or be diffused exponentially. At the same time, the rhythm of changes increases, followed by a quick change of values. In encounters with people from other groups or newly built networks, the values and norms become unpredictable.

1.2 The Impact of Individuation

The processes of survival and selection operate on the level of the individual. This is clearest in Darwin's favorite example: sexual selection (cf. Darwin [4]). Sexual partners either reject or accept one another, and the transfer of genes is blocked or enabled. This scenario presupposes *individuation* and a *context* (time, place, and the presence of other agents like rivals).

An individual is defined by the natural limits of life (i.e., birth and death) and the forces that sustain his/her life. If life is endangered, appropriate mechanisms exist to recover, repair damages, and avoid death. In this perspective, individuation has an implicit finality: survival under danger and risk. A tribe or family group may disappear if its reproduction rate is too low. Thus, in small and separated groups, a scenario characteristic of Neanderthals, the loss of some younger females may lead the group to extinction; if this often occurs and becomes a trend, the species may be extinct after several generations (particularly, if the overall population is small; it is assumed that the population of Neanderthals did not exceed 10,000 individuals).

By these dynamics, individuation can drive the evolutionary process, i.e., avoid or allow the extinction of a group or, ultimately, a species. Suppose individuals and groups of one species or subspecies share the ecology with others. In that case, their success may enhance the extinction of the other species if the mixture is genetically or behaviorally excluded or rare (dangerous).[8] Individuation is also the bridge for either biological or cultural innovation. For biological innovation, the individual that shows new features due to some biological change must first survive in the environment of individuals without this change and then produce progeny. These conditions decide whether the new capacities of the individual survive and finally change the characteristics of a group or even a species. In the case of cultural innovation, other individuals must perceive it as positive: imitate/learn it to become part of the cultural heritage. In each case, the change must be perceived and evaluated to trigger a new behavior, i.e., the effect relies on semiotic (sign-related) factors. Symbolic behavior evolution is enhanced by individuation processes and by the propagation and socialization of semiotic innovations.

1.3 Tradigenetic and Ratiogenetic Processes

One can distinguish tradigenetic and ratiogenetic processes (cf. Tembrock [17]). In the first case, sociocultural values, standards, and techniques are transmitted and conserved within a social group (e.g., a family, tribe, or a geographically connected set of tribes); the members of a group establish a tradition as a system of habits. Ratiogenetic processes enhance such events by distinguishing individuals or professional groups for their directive function. These persons may be older adults, priests (shamans), elected chiefs, etc. They incorporate the group's heritage (in their life memory) and can plan and direct specific innovations or dramatic changes that deeply influence a population's fate. Thus, the figure of Moses, who led the people of Israel out of Egypt, the prophets and founders of religion, or significant statesmen (see Alexander, Augustus in antiquity) and their helpers (a small subpart of the society) can rationally move a given society into a specific direction. Since the industrial

[8] Thus, Neanderthal individuals seem to have mixed with individuals of the early Homo sapiens when they met around 100,000 BP. However, a mixture in Europe after 40,000 BP, when Homo sapiens migrated massively into Europe, was rare. It did not prevent the extinction of the Neanderthals (possibly their offspring had health problems and rarely survived).

revolution (England 1760–1840), the French revolution (1799), and socialism (nineteenth century), technological, scientific, and political innovations have influenced the development of humanity.[9] Insofar as ratiogenetic processes refer to individual minds, i.e., the imagination and planning of individuals, they are part of individuation. In the transmission process, the traces of this origin may disappear because many people have contributed to its elaboration, transmission, and the establishment of new standards.

1.4 Morphogenesis and Mathematics for the Human Sciences

Christopher Zeeman and his disciples (mainly Ian Stewart and Tim Poston) issued a large number of applications ranging from models of heart rhythm, the behavior of dogs (flight/attack), and the dynamics of the human brain to social behavior (cf. Zeeman [25] and Poston and Stewart [16]). Consequently, qualitative dynamics (already sketched by Poincaré, 1874–1912) became an object of general concern. American and Russian mathematicians (Mather, Milnor, Smale, and Arnold) have further advanced the theory of singularities. Still, Thom's primary aim was to unfold further the morphogenetic intuitions he had described in his book *Stabilité structurelle et morphogenèse* [19]. Thom's and Zeeman's proposals have shown that the mathematical results in singularity theory and dynamic systems theory not only have deep historical sources in mathematics but are very promising for future research. Thom took up Aristotle's idea of "genus" (type) and introduced the terms "prégnance" and "saillance" which make up the heart of his "Semiophysics" (cf. Petitot [14], Thom [20], and Wildgen 2010b). Similar developments were at the heart of Hermann Haken's "Synergetics" and the models of the neuroscientist Scott [10], *Dynamic patterns: the self-organization of brain and behavior*). The shared strategy of these groups may be called "From matter to mind", i.e., the theories of the human mind are systematically founded on biological and physical systems theory and focus on the continuity between the natural and the human sciences.

René Thom, the famous mathematician, was indirectly attracted by semiotics and linguistics. At a point in his career as a mathematician, appointed to the Institute of Advanced Research in the Sciences (Institut des Hautes Études Scientifiques) in Paris, he began to reflect on the role of topology and topological dynamics in the fields of biology and the human sciences. In the sixties, Thom exchanged letters with C. H. Waddington, who wrote a preface to Thom's book *Stabilité structurelle et Morphogenèse* [19]. In the foreword to its translation into English (1975), Waddington refers to his book *Organizers and Genes* (cf. Waddington [21] and [22]), where he formulated some fundamental questions of theoretical biology, to which

[9] They even impacted the earth's climatic and geological state (cf. the exploitation of resources deep underground and in the oceans and changes in the chemical structure of the atmosphere and climate).

Thom found a mathematical answer. One problem with biology is that it cannot be explicitly founded on the laws of physics or chemistry. Such a foundation would ask for systems with millions or billions of factors. A topological treatment allows us to formulate the general contours of an explanation. In a later phase (prepared by articles after 1978), Thom wrote his book *Esquisse d'une sémiophysique* (Sketch of a Semiophysics, published in 1988 and translated to English in 1989). Thom tried to link the forces of the morphogenesis of meaning (semiogenesis) to primary magnitudes known from physics, i.e., gravitation and radiation (light). These universal fields embed the living beings and govern their environment (ecology). They are naturally the background of all perceptual and motor processes. In perception, light is at the basis of our visual perception; gravitation underlies human and animal motor processes and the sensation of pressure and weight; the ear registers sound waves, and the diffusion of chemical substances evokes reactions of our taste and smell organs.[10]

As the dynamics of such fields (e.g., light) have been the topics of physics since Newton and wave dynamics the subject of specific mathematical treatments since Maxwell, it was apparent for Thom to postulate a particular field registered and filtered by our sensory organs. He called it "saillance". The psychophysical field selects these effects as the most informative. Thom's program was to extract as much systematic content as possible from the analogy between physical and perceptual fields. Perception is the primary stratum of semiosis, and any perceptually based symbolic structure elaborates it. Although this strategy from physics to semiotics allows for the transfer of many mathematical techniques, there remains a large gap between psychophysics (the level of perception) and linguistics (or cultural semiotics). The term "prégnance" had to fill this gap and explain the transition from perceptual reactions to symbolic forms (culture).[11]

The dichotomy introduced by Thom has a precursor in Aristotelian philosophy. The corresponding notions for this duality are dynamis (δύναμις) or energeia (ἐνέργεια) versus morphe (μορφή) and typos (τύπος). Moreover, Thom refers explicitly to several modern scientific currents which contain a similar dichotomy:

1. In Pavlov's experiments, dogs salivate if presented with meat and "learn" the conditioned reflex of an associated bell (which evokes similar reactions). Here, the concept is used implicitly; the meat and the bell are "salient". The meat is "pregnant" by the biological constitution of the dog, whose body shows a set of automatic reactions, which are thus the content of the presented sign "meat". The

[10] After World War III, digital computers began to dominate the field of models in the life sciences and supplanted analog models. This trend is currently reversed as the digital models approach complexity limits. Analog models that use the resources of natural resources in physics, chemistry, and biology come to the fore. Cf. Dillavou [5].

[11] In linguistics, many responses to Thom's proposals were either simple repetitions or purely meta-theoretical. However, Thom's ideas were rather meant as stimulation for further research (comparable to his conjecture of a classification of unfolding dynamical systems that led to the mathematically elaborated classification theorem of singularity theory). After a phase of vague acceptance, theoretical linguists returned to the fleshpots of structuralism and its phenomenological epistemology. Cf. the comments on catastrophe theory in Piotrowski and Visetti ([15]: 25f).

associated sign "bell ringing" inherits the meaning of the sign "meat", and classical stimulus–response theory postulates that all meanings follow by stimulus conditioning from basic reflexes (elaborated by "operant conditioning," which starts from chance reactions, reinforced by success).

2. Konrad Lorenz observed the process called "Prägung" (imprinting) in birds (e.g., geese). During a short period after they leave the egg, many birds select rather unspecific stimuli in their environment and quickly elaborate on them to form basic concepts like that of a "mother bird". Thus, if Lorenz quickly uttered sounds and the freshly hatched goose registered his presence, he filled the slot of a "mother goose". As soon as this concept was established, he remained the prototype of a "mother goose" for them. Lorenz generalized this observation and postulated so-called "super releasers", i.e., very primitive schemata which only ask for minimal perceptual and neural control. They may even fit better into abstract molds than biologically real entities. In the process of "Prägung", a rich semantic system is developed which could not have been coded genetically.
3. Another biologist, Jakob von Uexküll, proposed a similar concept of "Bedeutungswelt" (meaningful universe). His theoretical biology influenced Cassirer's philosophy of symbolic forms when they were both lecturing in Hamburg. For Uexküll, every animal creates its own "Bedeutungswelt", which depends first on its windows of perception and then on its vital needs. Thus, a tick reacts perceptually to the concentration of butyric acid, typical of warm-blooded animals. The tick drops if this stimulus is received and eventually enters the animal's fur. This minimal semiotic system guarantees survival as the tick feeds on the animal's blood.
4. In psychology, Gibson elaborated the concepts of "prägnante Gestalt" and "Valenz" of his teacher Koffka and called it *affordance.* Any object or process in our environment may have affordances; thus, a chair allows for sitting, a bed for sleeping, etc. Therefore, the whole ecology is meaningful regarding possible actions or events it will enable. As some activities are more frequent or have more value for survival than others, the concept of affordance allows for grades. Maximal affordance would then correspond to "Prägnanz".
5. Cassirer generalized the perceptual/motoric "Prägnanz" to "symbolische Prägnanz" (symbolic pregnancy) as perception is the first level of semiosis (=symbol creation) in his system. One of the criteria of good gestalt was meaningfulness in Gestalt psychology. For Cassirer, symbolic content is the precondition of accurate perception as elements of perceptual input have to be integrated into a system of perceptually based concepts created by the mind. The signs associated with these concepts allow the steady stream of consciousness to halt as soon as something stable catches our attention.
6. The American psychologist David Hebb, whose work has become fundamental for modern neuropsychology (cf. Hebb-assemblies; i.e., networks of neurons with a similar function), discusses different types of instincts and two models. The first model, called monogenetic, assumes one primary force field of biological motivation, which then separates into various subfields like hunger, thirst, sexual

desire, aggression, fear, intellectual curiosity, etc. The second assumes a topology of motivations with different centers (attractors); cf. Hebb ([8], pp. 246–247).

7. William Stern proposed applying the concept of "Prägung" (imprinting) to language acquisition. He recognized a goal-oriented, internally controlled process in language acquisition. Thomae linked this idea with the more general discussion in ethnology and psychoanalysis (Thomae [18], p. 244). Freud's theory of psycho-genital development assumes far-reaching consequences of fundamental processes in early childhood, which fix specific personality traits for a person's whole life. Thomae (ibid.) explains this effect via a kind of "canalization", which directs all further developments.

Summarizing this discussion which has occupied biologists and psychologists for more than one century,[12] one can assume on the one side that all animals have specific windows to their ecology and on the other side that a small set of motivations or vital force fields exist. However, the list of relevant basic types remains an open question. René Thom presupposes the existence of such an essential and elementary set, which shares the feature of structural stability and is independent of specific material contexts. This basic set is subdivided into "salience" and "pregnancy".

To these rather general ideas already found in the discourse of the scientific community, Thom adds two new concepts:

- The salience effect in perception may be linked via psychophysical laws to the dynamics of objective fields in physics and chemistry; i.e., perception refers in its principles to laws in natural science. In Thom's terms, there is an explanatory continuum between objective salience in physics and chemistry and subjective salience in cognition and semiosis.
- The topology of salient objects and events governs the flow of pregnancies via a process called "diffusion de prégnance" (channeling of pregnancy). As in Lorenz's case of "Prägung" (imprinting), the attractor landscape of pregnancy fills with actual or memorized salience effects. In this channeling, the multiple forms of (perceptual) salience are elaborated into rich and context-dependent fields of categorical perception and behavior.

The self-organization of the differentiated pregnancy effects creates a system of meanings and, thus, the prerequisite for a lexicon of perceptual entities. The crucial step concerns the stability and complexity of repeated transfers in channeling pregnancy. The symbolic form is detached from the salient objects it designates, i.e., the reference to the sign object becomes independent from the time and place of the perception of the corresponding entity. This effect has two consequences: What triggers appetite or avoidance, for example, can be broken down into many aspects, in which the types of usefulness or survival relevance differ. For instance, the community can categorize flora and fauna according to edibility, medicinal effects, and instrumental use. The rich encyclopedias for flora and fauna in many collecting and hunting societies show the outcome of this multiple division of what is vital for an

[12] Cf. For further details of this discussion Wildgen and Brandt [24] and the chapters in Wildgen and Plümacher ([23]; in German) and Wildgen and Brandt ([24]; in English).

ethnic group. Applied to visual art, the image culture of a community is particularly striking for the artists and their clients in this society and directs their particular attention.

Thom's hypotheses apply mathematical techniques and theorems to solve fundamental problems. Nevertheless, the psychophysical transition calls for applying the laws of physical dynamics, not only the mathematics used in this field. The authors Turvey, Kugler, Kelso, and others (cf. Kelso [10]) have elaborated on this thought. For instance, the diffusion of pregnancy may correspond to fluid dynamics and models of growth in space and time.

References

1. Archer, M.S. (ed): Late Modernity: Trajectories Towards Morphogenic Society [Social morphogenesis, volume II]. Cham, Springer (2014)
2. Bailleul, R., Manceau, M. , Touboul, J.: A "numerical evo-devo" synthesis for the identification of pattern-forming factors. Cells **9**(8) (2020). Download PMC7463486
3. Blount, Z.D.: A case study in evolutionary contingency. Stud. Hist. Philos. Biol. Biomed. Sci. **58**, 82–92 (2016)
4. Darwin, Ch.: The Descent of Man, and Selection in Relation to Sex. John Murray, London (first edition, 1871) (1874)
5. Dillavou, S. et al.: Demonstration of Decentralized, Physics-driven Learning. ArXiv: 2108.00275 (2022)
6. Eigen, M., Schuster, P.: The Hypercycle: A Principle of Natural Self-organization. Springer, Berlin (1979)
7. Gehring, W.J.: Wie Gene die Entwicklung steuern. Die Geschichte der Homeobox. Birkhäuser, Basel (2001)
8. Hebb, D.O.: A Textbook of Psychology (new edition 1966). Saunders, Philadelphia (1959/1966).
9. Hidalgo, M., Curantz, C., Manceau, M.: A conserved paint box underlies color pattern diversity in Estrildid finches. Biology, 20 February 2021. https://www.semanticscholar.org/paper/A-conserved-paint-box-underlies-color-pattern-in-Hidalgo-Curantz/51726ba6a25ccef20a9d14d57e042bb253f936c5 (2021)
10. Kelso, J.A.S.: Dynamic patterns. In: The Self-organization of Brain and Behavior MIT Press, Cambridge (Mass) (1995)
11. Longo, G., Montévil, M.: Perspectives on Organisms. Biological Time, Symmetries and Singularities (series: Lecture Notes in Morphogenesis). Springer, Heidelberg (2014)
12. Maturana, H., Varela, F.J.: Autopoiesis and Cognition: The Realization of the Living. D. Reidel Publishing Company, Dordrecht (1980)
13. Nüsslein-Vollard, Ch.: Das Werden des Lebens. Wie Gene die Entwicklung steuern. Beck, München (2004)
14. Petitot, J.: Physique du sens. De la théorie des singularités aux structures sémio-narratives. Éditions du CNRS, Paris (1992)
15. Piotrowski, D., Visetti, Y.M.: The game of complexity and linguistic theorization. In: La Mantia, F., Licata, I., Percoti, P. (eds.) Language in Complexity. The Emerging Meaning. Springer (2017)
16. Poston, T., Stewart, I.: Catastrophe Theory and its Applications. Pitman, Boston (1978)
17. Tembrock, G.: Ökosemiose. In: Posner, R., Robering, K., Sebeok, T.A. (eds.) A Handbook on the Sign-theoretical Foundation of Nature and Culture, pp. 571–591. de Gruyter, Berlin (2004)
18. Thomae, H.: Entwicklung und Prägung. In: Thomae, H. (ed.) Entwicklungspsychologie. Handbuch der Psychologie in 12 Bd, pp. 240–311. Hogrefe, Göttingen (1972)

19. Thom, R.: Stabilité structurelle et morphogenèse, Interéditions, Paris (English translation: Structural stability and morphogenesis. Benjamin, Reading, 1975) (1972)
20. Thom, R.: Esquisse d'une sémiophysique: physique aristotélienne et théorie des catastrophes. Interéditions, Paris [English translation: Semiophysics. A Sketch. Addison-Wesley, Boston 1990] (1988)
21. Waddington, C.H.: Organisers and Genes. Cambridge University Press, Cambridge (1940)
22. Waddington, C.H.: The character of biological form. In: Whyte, L.L. (ed.) Aspects of Form: A Symposium on Form in Nature and Art (1968)
23. Wildgen, W. and Plümacher M. (eds.): Prägnanter Inhalt - Prägnante Form. [Pregnant Content - Pregnant Form]. Zeitschrift für Semiotik, 31(1–2) (2009)
24. Wildgen, W., Brandt P. Aa. (eds.): Semiosis and Catastrophes. René Thom's Semiotic Heritage. Lang, Bern (2010)
25. Zeeman, Ch.: Catastrophe Theory. Selected Papers 1972–1977. Addison-Wesley. Cambridge (Mass) (1977)

Chapter 2
From Sensorial Capacities to Symbolic Forms (With Particular Reference to Odor and Color)

Abstract Symbolic forms exist in the transition zone between sensation and communication, where perception and social cognition represent intermediate stages. In this transition zone, two primary symbolic forms near perception/sensation, i.e., odor and color, may serve as examples. In the first step, we rehearse the sensualist position of Condillac and his followers; in the second step, we discuss fundamental issues in the perception of odors and colors and their communication. The notion of "symbolic form" was introduced by Cassirer in the three volumes of his "Philosophy of Symbolic Forms" (1923–1929) and elaborated in further contributions on art, ethics, and politics. It will be a recurrent fix point of our analysis.

2.1 Introduction: Sensation/Perception Versus Communication

John Locke (1631–1704), Étienne Bonnot de Condillac (1714–1780), and Johann Gottfried Herder (1744–1803) have made clear that empirical sciences such as psychology, sociology, and anthropology[1] are essential for the study of sensation/perception and communication. Different focuses concern human understanding (Locke), the relation between sensation and knowledge (Condillac), and the rise of human language and culture (Herder). In his "Traité des Sensations" (1754), Condillac presents the following thought experiment. A marble statue is successively endowed with sensual experience. Condillac begins with olfaction. The statue first *is* what it smells; second, it distinguishes pleasant and unpleasant odors. Finally, comparison and memory select and stabilize the attention movement after the stimulus's reception. The difference is one of passivity if motivated by a stimulus and activity if motivated by memory (cf. Condillac [5]: 49). With the two principles of

[1] Condillac is considered the founder of French psychology, John Locke is one of the founders of political science, and Herder has made relevant contributions to anthropology in the late eighteenth century; cf. Condilac [4] and Herder [9].

W. Wildgen, *Morphogenesis of Symbolic Forms: Meaning in Music, Art, Religion, and Language*, Lecture Notes in Morphogenesis,
https://doi.org/10.1007/978-3-031-25651-6_2

pleasure/aversion and memory/comparison, Condillac develops a system of ideas based on olfaction. Similarly, he treats hearing, taste, and finally, sight, consisting of light and color. They all allow for positive/negative evaluation, memory/comparison, and thus order. The idea of extension in space, solidity, and identity establishes the shared limit. The integration into a world of apperceived objects is due to touch. The transition from sensibility to meaning/sense is possible, first, due to our mind's evaluative and abstractive capacity, and second, by integrating different types of experience. The "sensualistic" position of Condillac was a sound synthesis of ideas put forward in the discussions of modernity (sixteenth to eighteenth century) but underestimated the problem's complexity. The composition of larger structures of words/imagistic units asks for large-scale organizational principles, which may be called the "architecture" of sense. Two directions can be distinguished:

- Bottom-up: Sensations produced by our sense organs, triggered by external events, are stabilized, combined, and abstracted; they are memorized and come to conscience and thus make up the operations of thinking. This view has two critical zones: the psychophysical transition in the sensory organs (and correlated neural areas) and the combination and abstraction of perceptual "atoms" (produced in the first transition).
- Top-down: Any perception, memory, imagination, or combination of ideas presupposes the mode of thinking and thus incorporates principles and axioms of mental processing. The psychophysical contact zone immediately selects those data which fit the cognitive processing (eliminating or forgetting everything else).
- Self-referential: The thinking is shaped by the processing of sensual data. Piaget's equilibrium of assimilation (top-down) and accommodation (bottom-up) clearly expresses this view, which was already contained in Locke's and Condillac's essays (1690 and 1746). Locke makes the first and Condillac the decisive second move by including the problem of signs (words) in the following sequence:

word → (a sign of an) **idea** → (a sign of) **external event**

Ideas are constructs at an intermediate level and not independent of words' stabilizing and organizing functions. Perception is itself a semiotic activity. Ernst Cassirer took up the eighteenth-century tradition (via his preferred authors Herder, Goethe, and Wilhelm von Humboldt). The "symbolic form" cuts the continuous flow of sensations, categorizes it for memory, and thus links individual perception to collective structures like language, myth, art, technical skill, and ethics.[2]

Despite these advances, the following significant problems could not be solved:

- How are meanings *grounded* in our (subconscious) sensory activities, which are adapted by evolution to selected features of our environment? The quality of this grounding (the "correct" selection, the stable transfer of relevant structural

[2] In Wildgen [13], Chap. 9, they are called "genres of collective symbolic forms".

relations up to the highest levels of cognition) is crucial for the functional fitness of the "animal symbolicum", as humans are called by Cassirer [3].[3]

- How can individual thinking (as *re*presentation) be economical and still represent external reality?

Condillac considered perception as the first semiotic level and signs as the necessary stage to stabilize and organize memory and thinking. Thus a straight line of progress in complexity is proposed:

- Psychophysical input to the senses.
- Perception as the formation of signs for memory and imagination (dream).
- Language as a stage of stabilization and necessary for the organization at a high cognitive level.
- (Scientific) knowledge based on a linguistically organized universe of discourse.

However, this line of progress is incomplete because a fundamental conflict exists between individual perception, action, and *social* communication. The first field has as its domain the individual brain (including the sensory organs) and the individual development (maturation and essential learning in a suitable environment). The second field concerns the adaptation to a culturally transmitted system of linguistic rules, social beliefs, knowledge, social action, and perception.

Although processes of learning and socialization link both domains, the fundamental mechanisms are different. The organization of the brain cannot be identified with the organization of a community, and communication underlies other conditions of coding, transmission, functional adequacy, and economy than that operative in sensory organs and cortical centers of perception and motor planning. The differences become relevant in the use of olfaction and color and, thus, the contribution of sensory experience to unfolding symbolic forms.

2.2 The Perception and Communication of Odors

Humans are constantly exposed to odors. They may react to them positively or negatively, and olfactory impressions may control their attention and memory. Nevertheless, the level of communication is only reached if someone needs to speak to somebody, ask somebody, warn somebody, or order something concerning an odor. If both the speaker and the audience are exposed to the same smell, it is enough to call the audience's attention to the scent and then perform a speech act referring to the (supposed) common perception. It is only if the odor is not present and must be *re*presented that the problem of proper categorization and characterization

[3] The grounding of symbolic forms in the physiology and the basic functions of perception is a candidate for the type of "instructional signaling" due to ancient genes in Evo-Devo-biology. Compared to symbolic forms, the physiologically established perceptual mechanisms are ancient, conserved patterns that may instruct the formation of symbolic patterns; cf. the remarks in Chap. 1 (Darwin revisited).

occurs. It presupposes a stable system of meanings grounded in similar olfactory experiences. In a community where this demand does not occur frequently, communication regarding odors will be very insecure, unstable, poor, or meaningless. Thus, in the extreme case, every participant in a conversation may be perfectly able to perceive and, if necessary, distinguish a large set of odors but unable to communicate these distinctions. The required ability concerns *first* the stable grounding in personal perceptual experience, *second* efficient coordination with the perceptual patterns in those with whom one wants to communicate, and *third* the invention and use of labels (lexical or periphrastic). Thus, it does not help to learn words for odors; even if these labels are individually grounded in olfactory experience, olfactory communication remains vague because no *coordination* has been achieved. In expert communication, participants may accomplish such a grounding in training situations and by explicit conventions on language use. The experimental research in the perception and communication of odor in Dubois [8] makes clear that the grounding of odors is difficult or even impossible because the chemical structure of odors is very complex, and behavioral reactions are very context-dependent. There is nothing that corresponds to the Munsell classification of colors in the domain of odors. Dubois ([8]: 203f) writes:

> The stimulations originate in the organic molecules that trigger the specific olfactory receptors of the nose, transferring olfactory information along neural pathways through the olfactory bulb all the way to the brain, but with some specificity that makes this sense different from the others. Olfaction actually differs from the other more "noble" sensibilities: for one thing, the stimulations are described, not in physical but in chemical terms, as odorant molecules are mainly of organic nature. Furthermore, the neurophysiological description also insists on the specificity of the neural structures responsible for their coding.

Condillac [5] mentioned the problem of organization of the field of odors and the fact that it depends on the conditions and contexts of memorization.[4] Thus, if one experiences a series of odors, e.g., of flowers, coffee, and a nearby factory, a network that links and partially mixes these impressions and their evaluative reactions is created in memory. As a result, the neighborhood of odors in time and space, the amount of attention paid to them, and the pleasure or displeasure associated with them change from one individual to the next. Possibly, it is even more different for individuals living in other places and social and cultural contexts.[5] Nevertheless, it seems possible for linguistic communities to use a set of commonly understood terms or expressions to communicate odor impressions.

The range of linguistic devices to express the sensation and evaluation of odors has been underestimated in the scientific literature. Recent empirical studies (cf. the report in Dubois [8]) show that the linguistic repertoire used is much richer if the

[4] Cf. Condillac [5]:81: "Prenons par exemple cette suite, jonquille, rose, violette. Dès que ces odeurs sont constamment liées dans cet ordre, une d'elles ne peut affecter son organe, qu'aussitôt la mémoire ne lui rappelle les autres dans le rapport où elles sont à l'odeur sentie".

[5] Dubois [8]: comments the history of research in odors as follows: "In short, odors as smells, scents and perfumes are loaded with ambivalent symbolic investments that prevent exploring them as if such a historical and social context did not exist, in contrast with vision and audition considered by scientific research as 'purest senses' in both epistemological and moral terms (Cox [6], p. 1)".

interviewers ask appropriate questions about the perception and communication of odors:

> To summarize, the (still largely incomplete) inventory of linguistic devices to account for 'odors' as a sensory experience, when extended beyond the search for 'basic terms', reveals that olfaction is not a 'mute' sense. The lack of lexical items in olfaction depends on sociocultural and linguistic constraints of 'talking about odors' rather than physiological constraints. A large diversity of morpho-syntactic, syntactic, and discursive devices is at work, differing from the (lexical) categories relevant to vision (Dubois [8], 229)

Insofar as olfaction is, to a large degree, subconscious, the adaptation or social coordination is difficult. Therefore, communication either uses the individual olfactory perceptions and associated reactions only to a minimal extent, or olfactory perceptions are inaccessible to communication. Consequently, the perception of odors only approaches the threshold toward symbolic forms if coordinated with linguistic categorization and memory.[6] Therefore, we shall not explicitly deal with odor under symbolic forms.

2.3 The Perception and Communication of Colors

Condillac [5] argues that the perception of colors alone does not constitute colored places, situations, motions, or colored objects; it just forms a continuum of colored surfaces limiting one or the other. Our memory allows us to see two or three colors and journey from one color to the other. Nevertheless, the perception of colors contributes to an idea of extension, a *space* of colors. He also points to the difference between a naive observer of a painting and the painter.

> A painter and I see all the parts of the painting, but whereas he extricates them quickly, I have so many difficulties discovering them that I have the impression that I find in any moment parts that I have not yet seen. (translation by the author of a passage in Condillac (1755/1821: 111).[7]

The astonishing differentiation of human color vision does not automatically entail a rich system of "ideas" of colored surfaces in our minds. Attention, memory, the organization of the color space (surface), and paths of visual attention are necessary to produce a selected set of "color ideas". Higher levels are only achieved if other types of sensibility are integrated and form an organization linked to objects, events, and actions.[8] As the divergence of color terminologies in the languages of the

---

[6] This barrier is clear for animals like apes, dogs, and fish. Social coordination can, in these cases, only be enabled by instincts or by very few learned reactions. Aspects of linguistic reference to odors in texts about perfumes are treated by Peter Holz (2005; in German).

[7] Translation by the author: Condillac, 1755/1821: 111. "Un peintre et moi nous voyons également toutes les parties d'un tableau: mais tandis qu'il les démêle rapidement, je les découvre avec tant de peine, qu'il me semble que je voie à chaque instant ce que je n'avais point encore vu".

[8] Condillac argues that the sensibility of touch can control this higher reorganization. We can generalize this claim given a synesthetic reorganization of particular sensibilities (the old idea of a

world shows, the need for a basic (stable) set of color terms is very variable between languages, even on a similar level of cultural development. Therefore, communication about colors is not a universal and central concern of linguistic communities. This restriction does not preclude that the cognitive relevance of color perception is very high in individual perception, i.e., perception and communication do not share the same functional pattern and do not respond to identical (or similar) needs. They follow their specific principles and laws.

In the following, I summarize the significant findings of the color project directed by Kay (cf. Sect. 2.1.1 in Wildgen [14], in German).[9] The starting point (in addition to historical approaches by Young, Helmholz, and Hering) are publications by DeValois and Jacobs [7] on color perception in monkeys of the Old and New World, especially in macaques and marmosets. A comparison between macaques and humans in tests distinguishing flickering light from three steady light sources with decreasing intensity (same wavelength) indicates that color receptivity is almost equal for both species (see DeValois and Jacobs [7]: 535). A distinction is made between "non-opponent" cells, which generally register the light intensity, and the "opponent" cells, which are relevant for color perception. Four types of opponent cells could be distinguished (ibid. 538):

- (+ R, −G): They are excited by *red* and inhibited by *green*;
- (+ Y, −B): They are excited by *yellow* and inhibited by *blue*;
- (+ G, −R): They are excited by *green* and inhibited by *red*;
- (+ B, −Y): They are excited by *blue* and inhibited by *yellow*.

In anthropological linguistics, Berlin and Kay [2] have examined the elementary color vocabularies of different language cultures and the color values assigned to color words. A comparison with the physiological data resulted in a differentiation hierarchy, i.e., if the degree of differentiation of color terminology increases in one language or the transition from one language to another, the options can be foreseen by color physiology.[10] To a certain extent, culture-invariant color vision determines the basic color lexicon. This result speaks for a weak cognitive determinism. Figure 2.1 shows the hierarchy found.

Rosch-Heider and Olivier [11] compared the color categorization of 40 Dani (New Guinea) speakers and 41 American speakers. The Dani language has only two color categories: *mili* (dark and cold colors) and mola (light and warm colors). As in Berlin, and Kay [2], the Munsell color chart serves as the basis. The memory tests showed significant differences, i.e., the average number of correct identifications was 7.7 for Dani speakers and 11.7 for Americans. The question was whether memory accuracy is related to the color lexicon, i.e., whether a more differentiated inventory enables a better memory of color words. For this purpose, the authors compared perceptually adjacent color values concerning the same or different designations in Dani or

'sensus communis'). Cf. for an elaboration of this point, Chap. 7: "Eight Theses for (or Against ?) a Semiotics of Colour" in Caliandro (2017).

[9] The author translated selected passages.

[10] Kay and others continued the color project. Cf. for the current stage: Kay and Maffi [10].

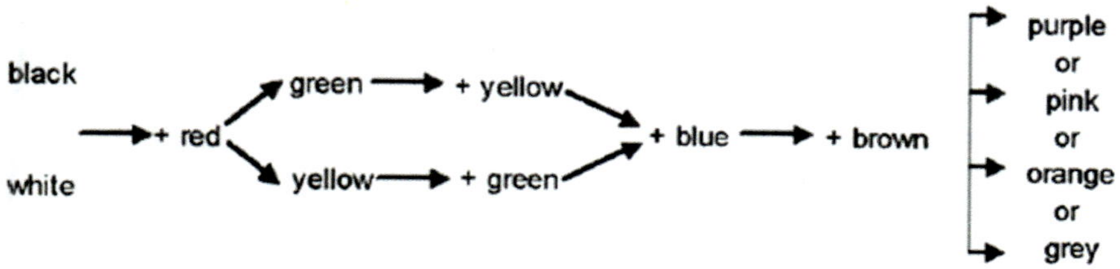

Fig. 2.1 Differentiation hierarchy according to Berlin and Kay [2]

American English. The differences were not significant,the authors concluded that the structure of the color lexicon does not significantly influence color memory, i.e., the cognitive performance of the color memory does not depend on the performance in the naming of colors. The authors also found that American speakers did not use basic color terms in almost half of the naming, preferring other strategies (object naming, metaphorical expressions, etc.).

Dubois [8] discusses the methods and results of the color project by Kay and others critically and adds research with a modified methodology on Palikur, a language spoken in Saint Georges de l'Oyapock, French Guyana. She sketches the method as follows:

> It (this study W.W.) confirms prior observations concerning the restrictive number of stable denominations of colors in Palikur, namely priye (black–dark), seye (white–light), duw˜e (red), ayeweye (green–blue), kwikwiye (yellow), that can be considered as "basic color terms" according to Berlin and Kay. But our procedure highlighted three other strategies mainly used by Palikur speakers when asked to describe colors and especially colors (equivalent of orange, brown, pink, violet, or grey) for which there is no simple lexicalization in Palikur (ibid. 103).

Dubois (ibid.) mentions the following strategies:

- A *morphological* strategy using a color term to which a shape classifier is added.
- A *syntactic* strategy where the color term comes along with a modifier (adverb or adjective).
- A *semantic* strategy of metonymy by using a word referring to another object to extract its color (ibid., shortened).

Although the universality of color terms or their sequence of appearance remains controversial, it became clear that putting the perception of colors (in context) in words or utterances is a complex activity that is, in many cases, successful. Therefore, skepticism regarding the linguistic realization of primary color perception is not warranted. Instead, color perception and the perception of spatial contours, lines, and other visual features can achieve the status of symbolic forms in a semiosis that uses more general resources. We shall discuss the symbolic forms based on visual perception in Chap. 4.

In a given linguistic community, some sub-communities, i.e., those of painters or other professionals, develop specific visual competencies and techniques of teaching or even arguing about colors which modify the demands on communication about colors. Models of colors like those designed for and by artists help to enrich and stabilize professional codes about colors and point to the underlying categorical architecture of communication.

Our reflections point to three significant problems or problematic transitions between "sensibility" (perception) and "sense" (meaning in a symbolic form, e.g., language):

(a) The transition from single sensibilities, e.g., olfaction or color sensibility, to an integrated perception is crucial.
(b) The transition between presentation and representation must consider the effect of attention, memory, and spontaneous imagination.
(c) The transition between perception, governed by principles of human neural architecture and dynamics, and communication, based on social interaction and organization principles, asks for a change in the theoretical frame.

The most dramatic transition and the evolutionarily most recent one is that of (b) to (c). At level (c), specific human capacities are concentrated in the plurality of "symbolic forms" discussed by Ernst Cassirer. They will be the topic of this book's Chaps. 3, 4, 5, and 6. As a kind of preparation for these central chapters, the notion of the symbolic form should be elucidated.

2.4 What Are Symbolic Forms?

Cassirer introduced the notion of symbolic form in his "Philosophy of Symbolic Forms", published in three volumes between 1923 and 1929. It contains two constituents: "symbolic" and "form". The term *form* was a central topic in Chap. 1 on morphogenesis, which includes the notions of morphè (form) and genesis (becoming, creation). The term *symbolic* refers to *symbol*, a polysemic notion used in philosophy and aesthetics. A discussion of the range of meanings of this notion would ask for an independent treatise. A proper starting point is the notion of symbol in the theory of signs proposed by Charles Sanders Peirce in the second half of the nineteenth century (cf. for further comments on Peirce and Cassirer Chap. 7).

Peirce considers three fundamental aspects of the sign, the central notion of semiotics, in its relation to the sign object, the content of the sign, and its reference to existing entities or processes in the world (the reality of dynamic processes): icon, index, and symbol. In this constellation, the notion of the *symbol* has to be delimited concerning the two neighboring concepts of the *index* (existential reference) and *icon* (reference via similarity, conceptual neighborhood). In a letter to Lady Welby on October 12, 1904, Peirce gives a summary of his position (developed after 1867, as he tells her; Wiener [12]: 391).

I define a Symbol as a sign which is determined by its dynamic object only in the sense that it will be so interpreted. It thus depends either upon a convention, a habit, or a natural disposition of its interpretant (that of which the interpretant is a determination). (ibid. 291f).

This definition needs clarification. The "dynamic object" is, in Peirce's terminology, the real-world object, the ultimate intention of the sign in its real-world usage. Peirce's statement tells us that the symbol only indirectly relates to it ("in the sense it is interpreted"). The interpretation depends on dispositions, habits, or conventions (see above). These determinations introduce a moment of arbitrariness; they depend on chance or, eventually, on many minimal causes beyond rational control. In the case of dispositions and habits mentioned by Peirce, one can assume rules of behavior that were acquired but gained law-like significance. In the case of conventions, these forces include cooperative effects in a community or conformity in social behavior.

A proper delimitation of the term "symbolic forms" (in the context of this book) says that they depend on the interpretation by individual users and conformity to the usage in communities. Their morphogenesis is, therefore, less natural or spontaneous than the morphogenesis in plants or animals. It presupposes the effects of human free will and the cooperative dynamics of socially conforming behavior. Previous sections of this chapter have shown that primary perceptual processes, such as those observable in olfaction and color vision, are still near to natural (bodily) morphogenesis. In contrast, colored surfaces and objects or artifacts (art) are symbols in the sense of the definition given by Peirce. In the case of odors, smell, and taste, artificial norms and devised terminologies can produce a symbolic level on which such perceptions are conventionalized and efficiently communicated. Nevertheless, we only treat music, art, language and myth/religion as fundamental symbolic forms in this book. Our analysis could probably be expanded to technology, ethics, and other communication fields following the lines of investigation demonstrated in Chaps. 3, 4, 5, and 6.

The first two symbolic forms of music and visual artifacts have a dominant founding in specific physical conditions: auditory perception and the motoric capacities to produce music in the first case, and visual perception and motoric capabilities for creating visual artifacts on the other side. In the case of myth and religion, a complex of bodily conditions, mental dispositions, and social demands shapes and changes the symbolic forms. Language, an evolutionary late-comer, has a broad field of interacting capacities and needs as sources. Its relation to myth and religion is one of co-evolution and interdependence. The symbolic forms of music and visual forms make use of the complexities of language to elaborate their repertoire culturally.[11] The multimodality of visual, acoustic, and linguistic communication is also a concern in social semiotics based on the linguistics of Halliday and further developed by van Leeuwen, Kress, Bateman, Wildfeuer, and others. See for an overview focusing on applications in web design, comics, film, audio-visual materials, and video games, Bateman et al. [1].

[11] A summary of the semiotics of the symbolic forms of music, visual art, and language in preparation for this book has been published (in English) in Wildgen [15].

References

1. Bateman, J.A., Wildfeuer, J., Hiippala, T.: Multimodality Foundations, Research, and Analysis: A Problem-oriented Introduction. De Gruyter Mouton, Berlin (2017)
2. Berlin, B., Kay, P.: Basic Color Terms: Their Universality and Evolution. Berkeley U.P, Berkeley (1969)
3. Cassirer, E.: The Myth of the State. Yale U.P., New Haven (1946/1974)
4. Condillac, E.: Essai sur l'origine des connaissances humaines. Editions Galilée, Auvers-sur-Oise (1746/1973)
5. Condillac, E.: Traités des Sensations. In: Idem Œuvres complètes, 1821–1822, vol. III. Reprinted by Slatkine Reprints, Facsimile, Geneva (1970). https://gallica.bnf.fr/ark:/12148/bpt6k96333766/f12.item.texteImage (1754/1970)
6. Cox, R.: Anthropology of senses. Int. Encycl. Anthropol. 1–12 (2018)
7. De Valois, R.L., Jacobs, G.H.: Primate color vision. Science **102**, 533–540 (1968)
8. Dubois, D.: Sensory Experiences: Exploring Meaning and the Senses. Series: Converging evidence in language and communication research; vol. 24. Benjamins, Amsterdam (2021)
9. Herder, J.G.: Ideen zur Philosophie der Geschichte der Menschheit. In: Kurz, H. (ed.) Herders Werke, vol. 3. Bibliographisches Institut, Leipzig (1784)
10. Kay, P, Maffi, L.: Number of basic colour categories. In: Dryer, M., Haspelmath, M. (eds.) The World Atlas of Language Structures (2013). https://wals.info/chapter/133
11. Rosch-Heider, E., Olivier, D.C.: The structure of the color space in naming and memory for two languages. Cogn. Psychol. **3**, 337–354 (1972)
12. Wiener, Ph.P. (ed): Charles Sanders Peirce. Selected Writings, chapter 24, pp. 380–432. Dover Publications, New York (1958)
13. Wildgen, W.: The Evolution of Human Languages. Scenarios, Principles, and Cultural Dynamics, Benjamins, Amsterdam (2004)
14. Wildgen, W.: Kognitive Grammatik. Klassische Paradigmen und neue Perspektiven. de Gruyter, Berlin (2008)
15. Wildgen, W.: The dynamics of human symbolic behavior: language, visual art, and music. In: Lear, A., Street, M. (eds.) Art and music. Past, Present, and Future Perspectives, chapter 2, pp. 32–64. Nova Publishers, New York (2018). http://www.fb10.uni-bremen.de/homepages/wildgen/pdf/THE_DYNAMICS_OF_HUMAN_SYMBOLIC_BEHAVIOR.pdf

Chapter 3
The Morphogenesis of Symbolic Forms in Music

Abstract The symbolic form of music was not in Cassirer's original list, but it is fundamental because it shows the impact of sensory and other bodily conditions on symbolic forms. The morphogenesis is first analyzed concerning evolution, starting with Paleolithic musical instruments (35,000 BP). The perception of music is the first dimension of embodiment. We can distinguish the organ of balance and the cochlea. Complex musical forms demonstrate the elaboration and diversification of basic morphogenetic patterns. Motion control and motion perception is the second relevant dimension in the morphogenesis of music. Finally, since baroque theories of music, emotion is considered fundamental. We apply an actual, evolutionary, and cognitive model of human (and animal) emotion. A particular focus is on melodies (or themes in larger pieces of music). Eventually, five applications range from songwriters (Bob Dylan and Leonard Cohen), a string quartet by W. A. Mozart, fugues by J. S. Bach, and quasi-narrative patterns in film music to innovations in the twentieth and twenty-first centuries.

3.1 Introduction

The biological[1] morphogenesis of sound production with the aim of communication goes far beyond our species and concerns a large part of the animated world. In the context of human communication and culture, the proper reference is the morphogenesis of symbolic forms in music linked to the auditory capacities of humans and the types of acoustic communication involving the human voice or ear and other organs able to produce and interpret human sounds. This analysis can be enlarged to musical instruments, i.e., the technology adjacent to and derivative from human sound communication.

[1] BP = before present; label introduced 1955 for prehistoric dates. For historical dates BP = BC + 1955.

W. Wildgen, *Morphogenesis of Symbolic Forms: Meaning in Music, Art, Religion, and Language*, Lecture Notes in Morphogenesis,
https://doi.org/10.1007/978-3-031-25651-6_3

The semiogenesis of musical forms has an evolutionary, a historical (developmental), and a synchronic or ad hoc creative dimension. We shall start with the evolutionary dimension and the cognitive dispositions for music that follow from it, specifically the embodiment of music and its relation to the physics of sound. The fundamental problem of musical semiotics concerns the meaning of music. It is assessed independently from the realization of meaning in linguistic forms (cf. Chap. 6).

3.2 Evolutionary Morphogenesis of Musical Forms

The acoustic manifestations of all living beings in an area are a part of the complex animated and communicative phonosphere, which can be distinguished from natural sounds such as wind and weather, falling trees, and flowing water.[2] In the origin of vocalizations and music by hominins, for example, in the savannahs or at the edges of forests in South and East Africa, the sounds of prey and predators and their accompanying noises had vital importance. At an early stage, such as in the species Australopithecus, which had a hard time with predators (such as leopards), it is likely that the precise analysis of the acoustic input, i.e., the location of extraneous signals, dominated. A further condition of auditory communication is the focus of common perception because acoustic signals orient the collective perception toward a selected field. The warning call indicates danger, a disturbance in the ecology. The sexual partner's lure replaces or completes social functions like a scent mark. The phonetic profile also serves to identify individuals and thus enhances social bonds, for example, between mother and child. At a close range, in safety (in the tree or the cave), the voices and sounds uttered by group members were the essential components of the phonosphere. They could be used in various functions, such as mother–child communication, sexual attachment, and tool-making/tool-usage teaching. Enhanced by these uses, acoustic signs gave relevance or meaning to environmental phenomena.

For the further development of music, the use of instruments was essential. These can first be characterized by their material composition. Even today, instruments made of wood or metal can be distinguished in an orchestra. Since metals were available as material only beginning with the Chalcolithic (Copper Age), wood, tree bark, reeds, grasses, and others were probably the basic materials. In addition, bones and horn components served as musical instruments. The earliest finds of musical

[2] The evolution of the larynx that connects the throat to the rest of the respiratory system is more or less contemporary with the evolution of vertebrates, and 230 million years ago, a broad spectrum of acoustic performances existed already. The families of animals developed different organs for sound production and reception on different scales of frequency. The physiological conditions for acoustic sound production near to language were fulfilled in the era of Homo erectus (1,8 million years BP). The question remains open at what date humans were able to use the symbolic potential of this acoustic capacity (cf. Habib [13]). Insofar as music does in general not presuppose this elaboration, it may be much older. Analogies of human music with the sound production of birds suggest a very deep evolutionary source.

instruments are those of swan bones (with perforations for sound modulation, age about 35,000 years BP), i.e., an archaic flute.

As African and Australian music traditions show, wooden instruments, e.g., drums made of hollow trunks, played a significant role in the development of instrumental music. These were supplemented by flutes (for example, based on a reed, bamboo, and animal bones) and simple string instruments. Examples are the instrumentation of bands in West Africa, which were the starting point for early blues orchestras: banjo, fiddle, and drum; the banjo combines a drum and a string instrument.

The masterful playing of these instruments could only succeed based on ethno-musical and ethnic-acoustic knowledge. Viewed from a modern perspective, the musicians have developed an intuitive, trial-and-error-based "technique" of music that has been handed down through generations. As lithic, wood, and metal technologies are a defining feature of our species, a close relationship between music and craft/technology is apparent. Additionally, both the mythical figures of Orpheus and the role of music in rituals (Olympic Games, Dionysus cult, tragedies of Sophocles) refer to the magical–religious and thus to the social and political dimensions of music. These usage contexts shaped the semantics and pragmatics of musical signs. A further important feature was virtuosity, which appeared in mythical figures, such as Linos and Marsyas, whose art aroused the envy of Apollo.

Within the human phonosphere, language has to be separated as a particular form from music, although there are numerous transitional forms (song, poetry) and cross-relationships (accent, rhythm, tonal languages). In the further development of human musical competence, the construction of instruments and instrumental music (drum, lyre, flute, etc.) played an important role. The separation of instrumental music from song and language since the seventeenth century is still characteristic of Western music and influenced globalized music traditions.

Human sound perception and production are fixed genus-specifically by the hearing and the voice organs (from the larynx to the lip). The conditions and potentials for song and speech are essentially the same. However, priorities are set in the respective cultural exploitation of the potential, or development opportunities may remain unused.[3] Instrumental sound generation mimics the physical condition of bodily sound production, extends it, and brings technical innovations into play. Thus, drums follow from stamping and clapping; string instruments and flutes imitate vocal articulations. In some cultures, inharmonic tones are used, equivalents of the consonants in human language (this also applies to modern music with noise components). The intensity profiles on the frequency scale, which are the basis for both vowels and voiced consonants, are, at the same time, the matter from which musical scales emerge (see Leman [22]: 63 ff.). Prosodic patterns of language, as well as musical beat, tempo, and phrase, make use of temporal characteristics. Together they form the shared basis of language and music in humans. This aspect will be elaborated on in Sect. 3.7.

---

[3] Cf. Also van Leeuwen [39].

3.3 The Embodiment of Music and Ecological Semantics

The analyses of meaning in language and other symbolic forms,[4] e.g., music and visual art, have evolved the last decennia from descriptive and logical strategies to consideration of cognitive aspects and aspects of embodiment, i.e., the whole body and not just the brain as relevant background. However, the human body (and brain) is in an evolutionary continuity with the bodies of other animals and dependent on the ecology, climate, and immediate context. The meanings of utterances, music, or visual designs go beyond arbitrary attributions of contents. The physical world, the body/the brain, and symbolic forms like language and music stand in the continuity of cause and effect. Scientific theories in this field should be able to cover this continuum. The French mathematician and philosopher René Thom proposed "semiophysics" as the proper designation of this endeavor.[5] Our analysis chooses a strict naturalizing strategy, i.e., it considers with preference aspects of musical meaning that can be connected to natural phenomena lying outside the proper domain of music and thus are promising candidates for explaining the musical meaning. Three foundational fields must be considered: first, the transition from musical aesthesis (sensitivity to music) to musical perception and performance; second, archetypal emotions as primary motivations of musical meaning and musical communication; and third, bodily motion and temporal patterns of music (linked to physical kinematics).

3.3.1 The Perceptual Sources of Symbolic Forms in Music

To find the meaning of musical signs, we must first ask for the fundamental meanings of acoustic signs in the context of human hearing. Since this was optimized in evolutionary history in specific directions, questions of the function and the biological impact of hearing arise. In the evolution of hearing, two lines are to be distinguished. They are manifested in the construction of the human inner ear:

- The balance organ is primarily responsible for the adjustment to gravity (top–bottom) and also corresponds to frequencies from 100 to 500 Hz at a sound pressure of 70 to 80 decibels. Deep and loud sounds and prominent rhythms, such as in dance music, lead to "hearing" with this organ.

---

[4] The following passages reconsider passages of the contribution of the author to the 15th International Congress on Musical Signification, in Barcelona, 15 to 19 of June 2022; cf. Wildgen [48].

[5] Thom's semiophysics Thom [38] can be understood in the tradition of psychophysics (e.g., Fechner and Helmholz in the nineteenth century). Parncutt ([33]: 16f.) even points to reflections of Jean-Philippe Rameau's"Traité de l'harmonie" (1721) that point in this direction. Parncutt coined the term "psychoacoustics": "*Psychoacoustics* is concerned with the relationship between the physical properties of stimuli and observers' descriptions of the sensations they evoke" (Parncutt [33]: 17).

- The cochlea was developed in land-living animals about 380 million years ago in the Devon period. As a result, it has a broad spectrum of hearing in humans and very high sensitivity.

The bourdon produces relatively static, mostly deep sounds, which serve as a reference point for the melody line. They are often built into the instrument, such as the bagpipe or the hurdy-gurdy. Mostly, they are tuned to the basic tone of a chord or the fifth. In Western music history, the basso continuo or ostinato played an essential role in the seventeenth and eighteenth centuries, and modern rock and pop music rely on very intense bass tones. It is this component of music that strongly motivates body movements. This music component's importance is closely linked to whole-body responses in the individual and the collective attunement to a simple, repetitive motion stimulus. The musical sign refers to the physical/physiological effects on the individual body. Secondarily, it serves the coordination of movements in the group, for instance, in dance, at work, or in a battle.[6]

The cochlea (in the inner ear) and auditory stimulus processing are characterized by their frequency domain (20–20,000 Hz), their sensitivity, the rapid processing of very complex sounds, and polyphonic or polyrhythmic structures. This feature is a prerequisite for most cultures' rich musical traditions. The invariants behind this cultural wealth refer to physical and auditory acoustics laws. Consequently, aural perception and the ecology of hearing, the "affordances" in ecological psychology (cf. Gibson [11]), have been tuned to these laws.

The physics of overtones in the human voice and musical instruments remains an essential pillar of human perception and control of musical forms. The overtones define a space of simultaneity and have a regular physical (and physiological) equivalent.[7] At the same time, this paradigm can be stretched in time, i.e., the resonating overtones can be "spelled out" as elements of a melody. As a result, the syntagmatic (sequential) order inherits the natural ordering system of the overtones. The space of possible auditory distinctions is vast at this primitive level. Sundberg ([35]: 62) computes a maximum of 1400 pitch distinctions and 280 distinctions of loudness. This wealth amounts to a space of approximately 400,000 distinctions. Memory

[6] The Scottish bagpipe, for instance, was a typical accompanying instrument in a battle, like a drum in nineteenth-century continental armies. Indirectly, this component bundles the attention and thus reduces the general stimulus and information flow. In this function, it has the opposite effect compared to language.

[7] The human vocal tract roughly corresponds to an open tube on one side. Essential for the tube are, firstly, the vibration depending on the length of the tube, and secondly, the sound showing the simultaneous presence of harmonics, whose intensity, however, decreases with frequency. This condition results in the so-called natural tone series, based on a division of the tube length by the natural numbers 1, 2, 3, 4, 5, 6, 7, 8, etc. Intervals naturally become smaller and smaller until they reach the threshold of human interval perception and perceptual intensity. A natural tendency in cultural expression was a limitation of the admitted dividers. Nevertheless, in terms of number theory, no consistent transposable scale can be set up (cf. Parncutt [33]: Sect. 3.3). The tempered scale was first adopted for keyboard instruments (cf. Bach's *Wohltemperiertes Klavier, 1722*) and then generalized throughout Western and other musical cultures.

effects impose the first severe restriction because memory works like a hedge in the cerebral recognition and assembling of complexes of musical signs.

As a consequence of these conditions of sound perception in humans emerges the phenomenon of categorical perception, i.e., the transitions from continuous scales (in physics) to boundaries imposed on this continuum, together with a radical reduction of the space of choices in musical perception.[8] Musical perception does not only concern single musical sounds but simultaneous complexes (cords) and sequences (themes, melodies). The physical and physiological interaction of single sounds with their array of overtones produces auditory effects called "rough timbre" or "ugly combinations" (cf. Sundberg [35]: 70). These phenomena correspond to periodic, quasi-periodic, and chaotic dynamic patterns.[9] The consonance and dissonance are an immediate effect of these natural, non-conventional patterns of musical perception. The primary *gestalt* characteristics of musical perception emerge from the physical and physiological laws and their correlation due to evolutionary processes. The dynamics of aesthesis build the immediate ground of musical meaning and open the mind's window for it.

The syntagmatic (sequential) order can inherit the natural ordering system of the overtones. This effect manifests in the tonic scale steps (I–VII) and the tonic levels: Tonic (I), Dominant (V), and Subdominant (IV), which show up most clearly in the classical final sequence (cadence) I–IV–V. Further central definitions, such as the major–minor contrast, and generally the classical tonality follow from it. Insight into the conventionality of this system prepared its abolition in the so-called atonal or pan-tonal music of the twentieth century.

3.3.2 Basic Emotions and Paths in an Emotional Space as a Potential for Musical Meaning

The meaning of music refers to fundamental processes connected to the human body. These include bodily needs, desires, and aversions that can be summarized under "Emotion". Music and emotion have been linked since antiquity Aristotle's affect-doctrine identified *lust–sorrow–dislike* as a set of fundamental affects/emotions from which he derived eleven composite affect-types. This classification played an essential role in history; it was modified many times until the Renaissance. René Descartes (1596–1650) suggested a resonance model explaining the effect of music, where the music moves human emotions (*musica movet affectus*). According to this idea, the harmonies in music trigger corresponding movements of the soul. They consist

[8] This transition is the standard topic of catastrophe theory and is encountered on many levels of semiosis. Cf. Wildgen [41] for catastrophe theoretic semantics.

[9] Sundberg mentions that if one adds a bass tone, this may "mask these ugly combinations" (ibidem 70).

of "spirit" (breath) flows and their cerebral effects.[10] In baroque music theory, this teaching was translated into compositional techniques. Both the keys and the interval ratios were related to emotional content. Example: small second > plaintive; minor chord > mourning; large intervals, i.e., deviations from the unison or small intervals > violent movement of the spirits. The characteristic of major and minor was arithmetically established in a rationalistic fashion. The "perfect" major triad, such as C-E–G with the simple proportions 4: 5: 6, was characterized as funny, cheeky, and sublime, and the "imperfect" minor triad A-C-E with the proportions 10:12:15, on the other hand, as sad or tender. This tendency remained influential in musical practice until the nineteenth century (see Eicken [8]). According to the baroque theorists, music's effect on affects is based on several features: the interval relations in melody and harmony, tempo and measure, dynamics, sonority, and ornamentation, and finally, the characteristics of the instruments, which can stand for single affects. Until recently, the affects/emotions themselves could only be described intuitively, as there were no new insights based on observational or experimental data. This situation has changed with modern evolutionary and comparative neurology. Therefore, it is helpful to start with their findings. Panksepp and Biven [31]: 16 divide emotional processing into three levels:

(1) Lowest level. Primary emotional processes refer to instincts.
(2) Middle level. Secondary processes of emotional control. These are learned in early childhood but cannot displace the primary processes (conversely, learning depends on emotional enhancers or inhibitions).
(3) Upper level. Knowledge and consciousness-driven emotion. Feelings such as trust, pride, shame, and guilt are developed at the middle level and can be elaborated into an ethical–religious code.

The following mainly deals with the two basic levels of emotional meaning. However, the upper level remains relevant to music's overall impact and embedding in social discourses and contexts.

Based on neurological observations (brain scans, local stimulations) and experiments (mainly with animals), Panksepp and Biven [31] propose seven "ancestral passions", i.e., evolutionarily fundamental emotional reactions that produce emotional complexes at the middle level. They form the basic vocabulary and the deep structure of musical meanings:

1. A more general emotion/motivation is called SEEKING: "The SEEKING system wants to motivate her [the mother] to find food and shelter" (ibid. 25). It often forms the background for other emotions, so it is a kind of primal motivation.
2. A group of emotional processes that cause aggression, violence, and fear, on the one hand, and lust, on the other: ANGER, FEAR, and LUST. These can be associated with specific chemical carriers and can be medically influenced in therapy (unless an intense fixation on the middle and higher levels prevents this). For example, music can be violent and cause fear. However, it can also generate

[10] Descartes interpreted the nerves he had seen during resection of corpses as tubes of a pneumatic mechanism.

feelings of pleasure. That is why music is often described as a drug or employed in the context of drug use.
3. The emotional dimension of PANIC/GRIEF is evident in separation distress and is therefore closely linked to the CARE affect, caring for the offspring, and other related parties. Complex middle-level emotional entities such as LOVE combine SEEKING, LUST, and CARE emotions. In specific cultural contexts, love codes can emerge, which in turn take shape in music codes; see, for example, the medieval minnesong ("Minnesang") and its sociological analysis by Luhmann [26].
4. Finally, Panksepp and Biven [31] introduce the PLAY dimension, universal in young people and adolescents. This motor-driven emotion is usually neutral to dominance and aggression and is primarily fun. In addition, roles are systematically exchanged. The field of available behavior is extensively explored in the game,it is clearly expressed in the animal kingdom beyond humans. This dimension is the most important for musical meaning. However, it is already at the upper end of the instinct-oriented first level in the transition to learned behavior. The game is also the basis for social well-being and laughter, social contact, and the management of coordination or conflict. It is one of the starting points in the evolution of human language (see Wildgen [43]: chapter 3 for the role of laughter in the evolution of language).

Game motivation is sensitive to disturbances, e.g., risk (FEAR), aggression (RAGE), loss (PANIC/GRIEF), or shortage (hunger, thirst). One can conclude that music generally requires a stress-free situation or promotes its realization. As an acoustic play, it works for both actors and listeners. The latter may dance, move rhythmically, or, as visitors to a classical concert, only perform inner movements such as dreaming and imagination. Panksepp and Biven point to the function "experience-expectant", i.e., the participants are in a state of expectation of a (joyful, pleasurable) experience. They expect certain moves, are surprised when different ones surface, and build up new expectations (see ibid. 374 ff.).

As seen from the above, musical meanings can be interpreted as movements in an emotional space. However, in some cases, subspaces may also come into play. For example, a battle song or musical accompaniment in battle (e.g., in Napoleonic times) mainly activates the RAGE dimension (and possibly the enemy's FEAR dimension). In contrast, songs for children and lullabies are more likely to affect the CARE dimension.

The ethnic diversity of symbolic forms in music may hide, modify, or elaborate the basic emotions just classified. For example, musical practices studied in New Guinea (Kaluli group) "are oriented to making the listeners sad and nostalgic, organizing feelings of sadness and eventually moving them to tears" Finnigan ([10]: 356). Other traditions (northern Sierra Leone) combine sorrowful songs "with the enjoyment of responding, clapping, and dancing." (ibid.). In ethnic musical traditions, the style and dominant emotions are part of a code,the songs are "thoughtfully composed and deliberately rehearsed" (ibid.) to trigger emotional effects. Such symbolic forms of music are at the third level (knowledge and consciousness: see above). Their embodiment is indirect via learned (or at least reinforced) emotional content. In rituals, on the background of myth and religion, they become integral to a (local) culture.[11]

The PLAY dimension, mostly without specific goals or purposes, has a side effect. It blurs the relation between musical forms and particular emotions. The meaning of the music becomes non-representational. It cannot be described as a projection of musical features onto a set of emotional values. Moreover, the musician completes his musical performance with gestures and facial expressions and adds personal feelings that the audience may perceive. They interfere with the emotional effects planned by the composer.[12]

3.3.3 The Meaning of Complex Musical Creations in European Music

In Europe during the last three centuries, music culture experienced the most astonishing heights in parallel with the technical and scientific development (from Bach to Haydn/Mozart/Beethoven and the New Music of the twentieth century). At the same time, popular music captured all areas of life. As a result, music has become part of an all-embracing medium of visual, linguistic, and musical signs in opera, musicals, and film. However, this complexity, this wealth, and the manifold relationships between various media make it difficult to properly assess the specific meaning or even the individual or collective effect/relevance of music.

A musical performance, whether it takes five minutes or two hours, gives rise to a kind of phonosphere. This harmonious environment captivates the concertgoer and the attentive listener in front of his sound system. When Stravinsky develops a complex sound fabric in the "Sacre du Printemps", he expects musicians and listeners to connect to a full "spring awakening" experience. A phonosphere is created in which listeners empathize mutually. They can use it as orientation in their overall perception of the music. The development of European music since Richard Wagner shows that very different sound architectures can create an impressive phonosphere with musical instruments and vocalists. This effect is pertinent even for the experiment of twelve-tone music. Although it radically changed the conventions of classical composition, it was obliged to fill the natural phonosphere of human hearing with a wide variety of musical motion and tension. Even serial music and modern experimental music, e.g.,

[11] Coded emotions open the risk of deceit or faking. Even in the case of professional musicians, the question of honesty or fidelity to the emotions intended by the composer arises. If the composer lived in a different period or context than the performing artist, it does not seem easy to guarantee the fidelity of the interpretation. The artist will rather stick to the score, or fake emotions he/she thinks are acceptable to the public.

[12] The strategies of musical performers may be very different. For example, some piano performers prefer a neutral, quiet behavior (bodily motion, mimics); others present vivid changes in their body posture and their mimics. This body motion (or neutrality) can influence, enhance, complete, or disturb the perception of musical meaning.

by Iannis Xenakis (1922–2001) and György Ligeti (1923–2006), are not meaningless. The composers merely abandoned meanings that had become cliché and looked for new possibilities of expression. After a relatively short period of negation, a new realm of musical meaning was opened; Hiekel [18] speaks of a two-year tunnel and a kind of musical Dadaism. This development is apparent, for example, in works by Helmut Lachenmann (born 1935: "The girl with the matches", 1997) or Clemens Gnadenstätter (born 1966: "Music for solo flute" from 1991). In this connection, Hiekel (ibid. 18) quotes a statement by Roland Barthes: "Once you have chased meanings out of the house, they come in through the window".

Loss and creation of meaning stand in a dialectical relationship, i.e., typically, a succession of change and persistence prevails. Returns to earlier stages are possible, and longer phases of stability or even rigidity may exist. The scales, chord hierarchies, and counterpoint relationships may be changed or even replaced, but movement patterns, tensions, and their resolution remain. Attraction and repulsion, reduction, and over-stimulation are the constants behind the change. It is, therefore, suggestive to see the proper existence of meaning in these movements and tensions. In the following, this track will be pursued, starting from the listening individual and the cognitive processes in their mind/brain (cf. the notion of mind/brain in Panksepp and Biven [31]).

3.4 Motion and Forces as a Potential for Musical Semiogenesis

It is helpful to start from a set of levels to formulate a model of musical semiogenesis based on human motion and locomotion and the forces that direct them. A rough classification into a sub-conceptual, a conceptual, and a symbolic (linguistic) level is straightforward. The first level works automatically and is based on sensory and cortical mechanisms; the second requires mental work and learning processes and is, therefore, culture-dependent. The third brings advanced symbolic systems (similar to the grammar of a language) into play. The dynamics (forces and induced movements) are naturally different at the three levels distinguished in Chap. 1 from instinct-driven to rational semiogenesis.

Many authors in the tradition of Lakoff and Johnson (1980), who have tried to apply the methodology of cognitive semantics to music, argue that the connection is a kind of metaphor:

> Musical motion must be some metaphorical motion that takes place within a symbolic space. One purpose of this chapter is to explain the metaphoric structure and logic of such motion and to ask what this means for how we experience and understand music (Larson [21]: 63).

"*Metaphor*" is a traditional concept issued in antique rhetorics and poetics. This means it is a high-level concept for the analysis of artful uses of language and not able to clarify basic cognitive or even physiological processes. Human motion and locomotion are first linked to physical processes in space/time; second, they exhibit motor processes of the body controlled by cognitive dynamics and specified in basic learning processes. They are not high-level functions of human creativity or poetic refinement. Third, multimodal processes, for instance, dance, ballet, and dramatic performance on stage, induce or coordinate motion. Therefore, the relation of bodily motion to music is a direct one, accessible to causal analysis and not just some poetic phantasm. There is no plausible reason musical analysis should persist in the vague, metaphorical perspective, where everything is a metaphor for everything else. We shall concentrate in the following section on the more basic, quasi-automatic correlations between music and human motion.

3.4.1 Bodily Motion (Kinematics) as a Source/Origin[13] of Musical Semiogenesis

The processes of physical exercise, e.g., walking, jumping, and dancing, are subject to dynamic laws roughly analogous to physical processes such as free fall, pendulum motion, and power transmission of billiard balls, with specific restrictions. The human gait in walking and running is a major source of rhythm in music. It shows in multi-modal phenomena involving music like a march- or dance music. The double meter seems to dominate, but the triple meter is quite frequent.[14] Walking and dancing are subject to coordination dynamics, i.e., the motion of the right and the left leg are coordinated. In horses, the speed of locomotion changes the motion pattern abruptly. In a simple experiment, the in-parallel motion of fingers changes to antiparallel due to higher speed (Fig. 3.1).

Kelso [19] uses the model Kelso-Haken-Bunz in Synergetics to describe such coordination processes generalized for different levels "From Matter to Mind". In this line, musical rhythms can refer to coordination patterns in the auditory brain, among groups of different muscles, and finally between physical subsystems in a coordinated mode (Hermann Haken took the primary mechanism from Laser-physics; other classical mechanisms are coupled oscillators).[15]

A side effect of this correspondence is that a metronome can implement the rhythm technically. The gesture is a particular form of movement that emerges from the field of physical movement and contains an intentional momentum; see

[13] Bodily and behavioral characteristics found in fully unfolded forms can be projected back to initial conditions that were decisive for all further developments. The source or source domain designates the starting point of morphogenesis, in which basic topologies and mechanisms can be detected. They are responsible for many of the developments occurring later on; cf. Thom ([37]: 168–172).

[14] In social semiotics (cf. van Leeuwen [39]: 48f.), the double meter is associated with the meaning "collectivity" (cf. marching or dancing in groups), and the triple meter with "artificiality/nobility".

[15] Human motion was already a topic for Galilei, who is the founder of modern kinematics and dynamics. In Synergetics, an interdiscipline based on laser-physics and thermodynamics, the gaits of animals and the coordinated motion of humans are analyzed on par with other spatiotemporal patterns in the inanimate world; cf. Hafen ([14]: Part II: Behavior).

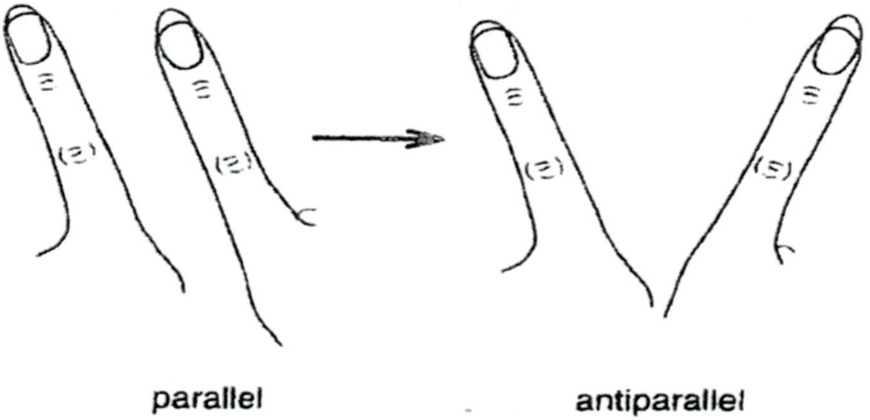

Fig. 3.1 The behavioral experiment of coordinated motion (cf. Haken [14]: 69)

nodding, pointing, or eye-gazing. In music, gestures have a relatively broad intention linked to internal states or movements. A melody may be decomposed into several gestures, some of which are within the boundaries of beats, some exceeding them. Kühl ([20]: 168 f) demonstrates a division into musical phrases/gestures based on examples from Mozart (Piano Sonata in A major, KV 331) and Charlie Parker (Bloomdido). The emotional effects of pitches and their movements (high–low tone, ascending–descending) have been frequently discussed but seem to be culturally and individually very variable (see ibid. 177 f.). In any case, the internal correlate is an emotional movement neurologically associated with a brain area, e.g., connected to the hypothalamus and the limbic system. These "movements" are correlated with the release of messenger substances and the activation of electrical signals.

The conceptual level (without symbol structure) can be described as a tonal space, so melodies are movements in this tonal space. Classical terms such as scales, keys (e.g., major and minor), chords, and tonic levels (I–VII) can be introduced here. The dynamics can be mapped on paths through the tonal space. Motion is described by vectors (direction, strength) and similar mathematical concepts of linear algebra. On the symbolic (quasi-linguistic) level, partial movements (melody phases, motifs, movements of a piece of music) are arranged into a discursive unit. Narrative and rhetorical patterns (metaphors and metonyms) will most likely become relevant.[16]

[16] The dynamics are similar. For an overview of cognitive grammar, see Wildgen [44]. Metaphors and metonymy are treated in the chapter: Cognitive models and metaphor (Lakoff), 65–90 (in German).

3.4.2 *Bodily Forces (Dynamics) Motivating Musical Semiogenesis*

Physical forces apply when a movement is redirected (change of direction), accelerated, or decelerated. Larson ([21]: 321) describes (ibid. 312) "musical meaning as an emergent property of the musical forces", but he understands "musical forces" in the sense of Talmy's "force dynamics". In our morphogenetic view, musical forces are embodied physical forces. Three types of musical phenomena can be linked to embodied forces:

- Forces are effective when a movement is redirected (change of direction), accelerated, or decelerated. This can be done in the three areas indicated above.
- The rhythm can be doubled or varied (e.g., from 3/4 to 2/4), speeded up, or slowed down. Messiaen proposed complex transformations of the rhythm in 1944. For example, the addition of an eighth note or the punctuation of a note produces the modified rhythms: 3/4 → 13/16 or 2/4 → 5/16 (see Bogue [5]: 26). The acceleration and deceleration are in an indexical relationship to physical and mental movements or affects; the audience is moved (in upward and downward movements), goes along, or is synchronized with the rhythm of the music. Often a balance is sought over a longer period; acceleration is followed by deceleration. This can also be observed in audio-visual art, for example, in film (see Wildgen [45]: Sect. 7.4: 217–233: in German; and Wildgen [46]: in English).

The melodic structure of a song can emanate from the main tonic level and end with it. In the intermediate field, dominant and subdominant transitional fields build a tension of difference. In piano playing, the two hands can have different melody lines, or one hand can have an accompanying and decorating function. The same applies to the choir (canon or polyphonic singing) and orchestral music. The art of counterpoint exploits the tension between the main motive (Dux and its variations) and the countermovement (Comes). The iconic (imaginistic) reference to gestures (e.g., visible in dance) implies a gestural dynamic consisting of accelerated/slowed gestures, stagnation, complication phases, and intensity maxima. The indexical (causal) relations correspond to vectorial relations, such as a force transmission (type billiard ball), addition, subtraction of vectors, and more complicated patterns. In this domain, concepts of linear algebra can be applied (instead of vague "metaphors" or diagrams in Lakoff's or Talmy's style).

The mathematician and physicist Leonhard Euler (1707–1783) dealt with the determination of dissonance/consonance as early as 1739, setting up a sequence of consonances: prime > fifth > fourth > major third/major sixth > minor third/minor sixth/great second > minor seventh > major seventh > minor second and the tritone (see Mazzola [27]: 57). In general, musical intervals can be described by a landscape of attractors (sinks of a potential). For example, Chella ([6]: 196) shows a dissonance curve for the scale values of C4 (roughly in the middle of the piano keyboard). The dissonance measure of the pair (C4, G4) is 0.39 (consonant), while the dissonance measure of the pair (C4, Db4) = 2.56 (dissonant). Taking the frequency scale as abscissa gives a potential curve (cf. Fig.10.7, ibid. 196 and Mazzola [27]: 61).

3.4.3 The Morphogenesis of Melodies

Melody and pitch movements are paths in a landscape[17] where the dissonance minima are attractors (here, the movement stops), dissonance maxima are avoided, and moderate dissonances serve as transition areas. These movements generate tensions, which can contribute to a specific musical meaning.

Melody and forces

Melody can be memorized, stimulate repetition, and, as a whole, justify appreciation (see the effect of catchy tunes). In more extensive pieces of music, the melody is usually limited to one phrase, one motif. Its character as a building block becomes apparent when the melody is repeated, varied, and transposed, i.e., presented in different "dresses". It then acts like a person, a role in a drama. Primary melody curves already show up in the infant's crying, which is the phonetic starting point of language (see Wermke and Mende [40]). The most straightforward building block is the rising and falling fundamental frequency (F0), ranging from 300 to 500 Hz,it forms a melody arc. This structure is further developed in babbling and later forms the underlying musical form of two-syllable words (see ibid. 637f.). However, the laryngeal control of phonation by the infant must first be expanded by the muscular control of the oral cavity so that more precise articulations can be mastered, which allows the distinction of phonemes and words. It is also the context for the unfolding of the musical melody. The vocal range of humans usually frames the melody of a song, e.g., the song corresponds to the comfortable tonal space of a male/female/child voice. This tonal space has an upper and lower edge and an intermediate zone. The melody can go from the bottom to the top, stay in the middle for some time, progress step by step, or move variably back and forth. In addition to a general up-and-down movement, one can also observe fast-changing variations of the melodic line, as they often occur in jazz variations. The experienced instrumentalist "decorates" the melody line with fast movements and trills.

The course of the melody has a center that can be used as a zero expectation of melody progression, i.e., no change of pitch or a minimal one results in no melody, a "unison". The melody is then the deformation of this standard, and the deformation energy can be determined from the deviations from the center. Most folk music and songwriters' songs have a flat melody curve that approximates that of language. The curve then descends toward the end of the line or stanza or increases (as in the

[17] The mathematical space of the 12 half-tones of the Western (tempered) scale can be constructed as a torus of tierces; pathways and symmetries on the torus (cf. Mazzola [27]) stand for basic movements and dynamics in music. In a similar way, rhythms and meters can be treated mathematically,cf. Papadopoulos [32].

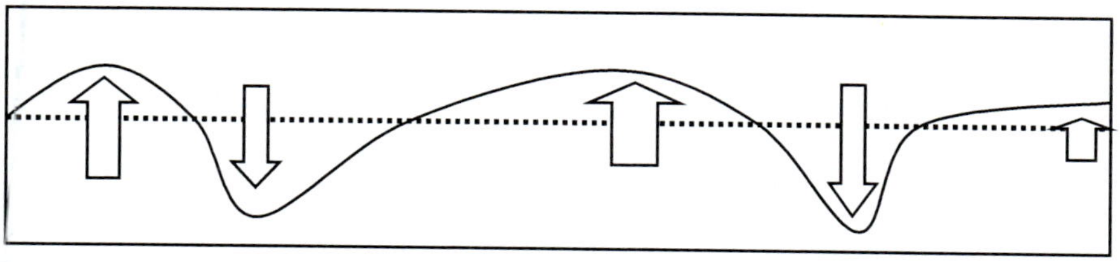

Fig. 3.2 The melody course as a result of forces acting on a medium level

intonation of questions). Leyton [23] assumes a "memory" of the deformations (of ideal forms) as the specific meaning of a melody. It contains information about forces that have become effective and influenced the melody's design. For the performer, for instance, the singer, these forces correspond to efforts that the voice must make to produce and control higher or lower notes. For the listener, either it is more difficult to perceive, recognize, and process high and low tones, or the listener imagines the forces applied as a potential performer. The professional opera singer may indulge in artistic coloratura and amaze his audience; the heroic bass may force his voice into the deepest cellar. However, we find very flat melodies when we look at the catchy hits or songs that delight a mass audience. The forces that shape the melody, moving it away from a neutral middle tone, can be imagined as the effect of forces that push the melodic line up or down concerning the middle (comfortable) level (Fig. 3.2).

Overall, a melody can thus be strong or weak concerning applied forces. Powerful and low-energy passages can alternate in a song or a piece of music. In the final result, it often comes to a balance, or either the ascending or the descending forces predominate. These movements are perceived as sign shapes and associated with a dynamic meaning. Downward movement may, e.g., express sadness or resignation, and an ascending movement, on the other hand, enthusiasm or hope.

Levels of motion in a melody

It is also possible to distinguish a "melody in steps" from a "melody in triads". In the first case, e.g., in Gregorian chant, the melody more or less follows the steps of the scale, in seconds and thirds, rising or falling. The singer's experience associates ascension with a (muscular) tension, an effort; descent to the center is experienced as relaxation or ease. The same applies to acceleration or deceleration of the tempo. The triadic melody uses the different tensions of the intervals in a melody. Seconds, thirds, fourths, fifths, sixths, and sevenths (intervals) have different tension values. The partial movements of the melody can also be assigned to a chord scheme, i.e., they realize in time the consonant notes of a chord or a dissonant combination of pitches. These movements and tensions (which refer to forces) thus represent the possible meanings ("signifié", significant) of a melody. The duration of individual notes also plays a crucial role. In the late Middle Ages (thirteenth and fourteenth centuries in Europe), the duration of the notes, which were initially based on the length of the syllables (long vs. short), was further differentiated. The mensural notation facilitated the fixation of this innovation. In the polyphonic movement, one voice could be noted in longer note values (for example, the bass), while the other voices used much smaller note values, that is, a faster melodic movement. The chords to which the steps of the melody refer are arranged in levels: tonic, dominant, and subdominant (in the context of tonal music). The melody can stay within one level, exploit the possibilities, and switch to another. This directly affects the accompaniment, which is why, for example, the guitar accompaniment for a song is indicated by the key above the text or the musical notation. In many cases, the notation of the keys is a kind of shorthand melody, i.e., the musician can remember the melody based on the chord progression and vary or embellish it within the given scheme. Important centers of attraction (tonal centers) and organization are.

Tonic level → Dominant level → Subdominant level

Within each level, the possible sequences through the scale and chord structure are specified as a triad, four-note chord, and others. The effect of the melody is based on the sequence of stages, on the possibility of a change from major to minor and vice versa, and on the tensions between the individual steps, which can also be varied through different durations, accents, and tempi. A song's relatively simple chord scheme may be elaborated in its further development (e.g., from a simple blues song to jazz interpretations or improvisations). Berendt ([4]: 132f.) shows this in the case of a popular jazz theme of the 1920s: "I cannot give you anything but love". The original chord progression: C6–Am–Dm–G7 is elaborated as follows: C6–Dm7–Em7–Ebdim–Dm7–F6–Dm7/9–Db7/9. In the further variation and improvisation of a theme, there may be variations in both the execution within a chord scheme and the sequence of chord schemes (within a particular harmonic framework). If the theme is generally known, it can even be pushed into the background; it remains implicitly present through individual features or characteristics (for the performers and competent listeners).

The meaning space of music

A central problem of dynamic modeling in musical semiogenesis concerns the nature of the space of meaning in which the performer and the listener move. We have already referred to the space of objective, physical motion as a tonal space, e.g., the frequency range of a voice or instrument. In the case of music, the space of meaning does not correlate with the space of the world as experienced visually or through action but instead with an inner space, which, however, must be transparent and comprehensible to musicians and listeners. One can also call it the mental space of music and then try to use neuroscientific methods to test hypotheses about the structure of this space. For the time being, it is a hypothetical construct based on the intuition of musicians and their listeners and the analysis of musical works. The temporal structure of the melody can be enriched or modified in three ways:

1. *Variation of the rhythm.* In the development of jazz, a variation of the original marching rhythm in the New Orleans style with an emphasis on the first and third beats in the 4/4 m was followed by an emphasis on the second and fourth

beats (in contrast to European tradition). This happened in the Dixieland and the Chicago style of jazz. In the bebop style, the staccato of the accents is replaced by legatos, creating a kind of continuity. Finally, fast and changing rhythms dissolve or override the fixed meter in free jazz. The dominance and freedom of rhythmic design is a trait that jazz has inherited from African traditions or rediscovered through a return to ethnic music cultures from different parts of the world.
2. *Adaptation to individual instruments and situational demands.* The artists design a melody regarding their instrument (for example, the saxophone, the clarinet, the trumpet, and the guitar). Additionally, they can adapt their performance to momentarily perceived internal dynamics, i.e., to the physiological–emotional component of their performance.
3. *Adapting to other musicians and the performance of the band.* The soloist "resonates" with the performance of others. The result is often referred to as *swing*; it captures the players and then the listeners. This aspect is especially pronounced in the religious gospel, where the (singing) preacher brings the church to ecstasy (see Halle [15]: 143–152).

This approximate analysis does not reveal the secrets of a perfect melody. What makes a melodic line/phrase so memorable that it becomes a catchy tune or circulates in countless versions in a musical culture for decades? How do the diverse melodic, harmonic, and rhythmic tensions combine to form a successful figure that gains cultural significance? At the basis of this problem, we suspect complex cognitive and social (group-specific) coordination processes in the interaction of a band with its audience. The underlying cerebral and communicative processes are still largely unknown.

After this presentation of rather general, dynamical concepts for the morphogenesis of music, more specific features of musical composition will be analyzed. Our starting point is songwriters' music, specifically Bob Dylan (1941–) and Leonard Cohen (1934–2016). A further section exemplifies aspects of the PLAY dimension of music in the string quartet, understood as a conversation between four instruments (and players). It is exemplified with the help of the String Quartet No. 19 in C Major, K. 465 by Wolfgang Amadeus Mozart. Next, complicated sequencing patterns in the fugues of Johann Sebastian Bach elaborate the PLAY dimension in musical morphogenesis. A further example concerns music's processual and quasi-narrative aspects (for instance, in program and film music). Eventually, more recent advances in muso-mathematics and computerized music are discussed.

3.5 Some Examples of a Morphogenetic Analysis of Music

The focus of our analysis on the morphogenetic aspects of music has the consequence that evolutionary, developmental, and historical processes are highlighted. This decision is a programmatic step that negates the neglect of all questions of becoming or causation linked to natural phenomena in the tradition of structuralism (after de

Saussure 1916 and followers in the twentieth century). Nevertheless, the morphogenetic analysis does not preclude the analysis of existing music works, their design, and their efficiency. On the contrary, the dynamic paradigm allows for new insight into classical musical analysis. Our analysis starts with popular music, specifically the music of songwriters Bob Dylan, the Nobel prize winner in 2016, and Leonard Cohen; see Wildgen ([47]: 125–140) for a detailed analysis (in German). The other examples focus on classical music, the string quartet (Mozart), the art of fugues (Bach), program music, quasi-narrative music features, and computerized music.

3.5.1 Gestures in Bob Dylan's Song: Tambourine Man (1964) and Musical Forces in the Song "Hallelujah" by Leonard Cohen (1984)

Musical gestures establish an external reference, i.e., the sphere of musical meanings goes beyond and "transcends" intra-body references and their empathic transfer to other bodies. A musical gesture can have the form of a short melody or a phrase in a larger piece of music. A higher level of organization is reached by "phrase groups". At this level, conflicts may appear. Drums or other rhythm instruments can express the dimension of a body-based meter, e.g., the guitar played by Bob Dylan in example 1 (cf. Table 3.1). His singing, however, depends on the phrase structure of the text. In the coordination of both dimensions, a rich field of interactions is activated, which is further complicated by microstructures, i.e., minor deviations, often characteristic of the style or the performance of a singer/musician.[18] The morphogenesis of musical forms at this stage manifests principles of self-organization, coordination between different, even conflicting forces, and ratiogenetic procedures in optimizing the commercial product, public musical performance.

Table 3.1 Metrical and phrasal analysis of the beginning of Bob Dylan's "Hey Tambourine man"

Meter 1				2				3					4
G	G	A	A	D	D	G	G	D	D	G	G	A	A
Phrase 1				Phrase 2			Phrase 3		Phrase 4				
Hey Tambourine man				Play a song for me			I'm not sleepy		And there is no place I'm going to				

[18] Eventually, quasi-rhetoric patterns show up in the sequential schema of Exposition, Development, Recapitulation, and Coda. The art of sequential organization can be admired in Bach's work on fugues and the tradition of counterpoint. Although such sequential patterns may allude to syntax in human languages, a denotative, descriptive, narrative, or even argumentative organization of meanings and referential intentions is lacking. This is not a shortage but a virtue of musical communication. The complementary nature of musical versus linguistic meaning opens the field for enhancing interactions between language and music in multimodal performances.

The guitar of Bob Dylan plays the musical rhythm; the motions of his right hand realize the basic meter, his left-hand moves from one chord to the next, and his voice sings the melody and articulates the text. Thus, his performance realizes three types of bodily motion patterns. In a strict sense, only the third part can be analyzed as a sequence of (indexical) gestures, i.e., it points to a goal and has an intention, whereas the singing voice follows the phrasal structure of the text. At the end of the whole stanza, Dylan adds a 2/2 m (not shown in the table) to finish the stanza on the first meter of the next unit instead of the third beat of a 4/4 m (cf. Kühl [20]: 189).

The melody in the song "Hallelujah" by Leonard Cohen moves in a flat curve around G/A in the first ten beats. After "It goes like this", it climbs up in steps until it reaches the E', one octave above the starting point. The repeated "Hallelujah" descends in a wave-like movement until it reaches the lower and final C, the resting point (the song is written in C).

Professional singers can range up to two octaves. The usual range of a male singer is little more than one octave. Leonard Cohen starts on a deep level (like a bass); thus, he can master the range from the low C (the final note) to the high E without difficulty. For many lay singers, the range of this song asks for special effort. The central area between C (low) and E (high) is around G/A. This was also the dominant area in the introductory ten beats. If we apply Leyton's concept of memory of forces, we can draw the following schema of applied forces (Fig. 3.3) (Video link: http://www.leonardcohen.com/).

Leonard Cohen started his career as a writer of poems and turned to music (in an interview, he said that selling music allows for a living while selling poems does not). He wrote almost 40 stanzas of this song. In one of them, mainly the first one performed, he even described the musical movement of his melody:

It goes like this, the fourth, the fifth
The minor fall, the major lift
The baffled king composing Hallelujah

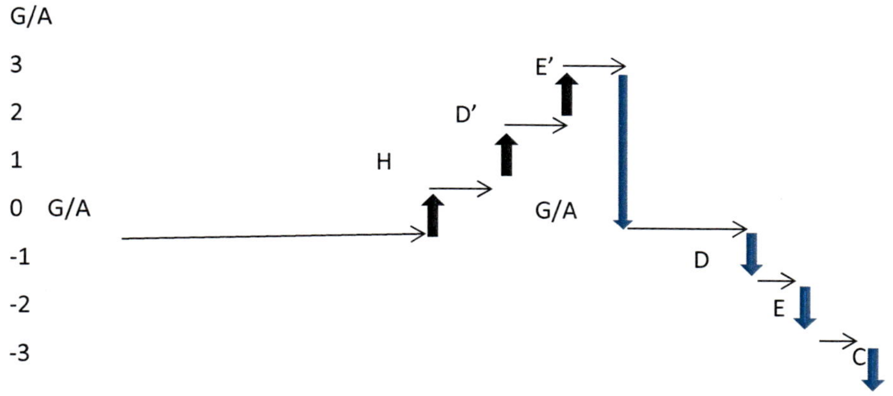

Fig. 3.3 Motion and force in the song "Hallelujah" by Leonard Cohen

Thus, the author of this song was aware that making music is a kind of itinerary in a musical landscape.

3.5.2 *The String Quartet as a Musical Conversation*

Complex musical forms have existed since ancient times, and the records of the music of non-European and remote ethnic groups attest to the wealth of musical inventions in all cultures. Polyphony is only one aspect of such complexity. Polyphonic music probably arose in free improvisation for the pleasure of sounding together. From the thirteenth century onwards, it was institutionally consolidated and written down. It reached its peak in the form of the string quartet, introduced by Luigi Boccherini and Joseph Haydn around 1760 and masterfully executed and further developed by Haydn in particular. The starting point was probably the trio sonata for two solo instruments and continuo that Arcangelo Corelli (1653–1713) had developed from 1681 onwards. The quartet eventually was the perfect form, for Haydn experimented with a quintet but returned to the quartet. But why four instruments from the violin family (two violins, viola, and cello), and why just four musicians? The violin family covers the required tonal range and allows for a great variety of expressions both through the variability of the individual instruments and their characteristics as through the expressiveness of the moving bow and the tremolo of the fingers (the strings can also be plucked). In addition, the number four has a discursive plausibility: four educated people exchange their opinions, respect each other in their diversity, and harmonize. It is not about the duel of two opponents but about what we have in common in diversity.[19]

This diversity and the specific role played by the violoncello are visible in the score of the “String Quartet No. 19 in C Major”, by Wolfgang Amadeus Mozart. Three voices start one after the other; the violoncello plays a simple bass pattern. All three instruments are present in the third bar and coordinate to a crescendo. The harmony in this phase is slightly dissonant, which is why this quartet has also been called “Dissonance”. In the Allegro part of this introduction, the dissonance is resolved in a C major. The music alludes to different cooperative (occasionally slightly disharmonious) conversations (Fig. 3.4).

The playing of the string quartet is also transparent to the listener; in particular, the individual instruments and their performers come to their right but with varying emphasis. With a large symphony orchestra, this transparency is more difficult to

[19] Mazzola [27] explains in the chapter "The Theory of the String Quartet" that the number of four instruments represents the minimum for the complete development of a three-dimensional tonal space (pitch, deployment time, duration) with two technical parameters (bow pressure, bow speed), i.e., in a five-dimensional parameter space. The harmony of triad coverings leads to a map with three elements multiplied by a three-part cadence, i.e., nine dimensions. The inequality gives the number of instruments required (n): $n + 5 \geq 9$. According to this, the number of instruments must be at least 4: $4 + 5 = 9$, which satisfies the inequality. The form of the overall musical sign, in this case, the concerto of the string quartet, is based, at least in its contours, on the formal conditions resulting from the material (the instruments) and the tonal space to be filled.

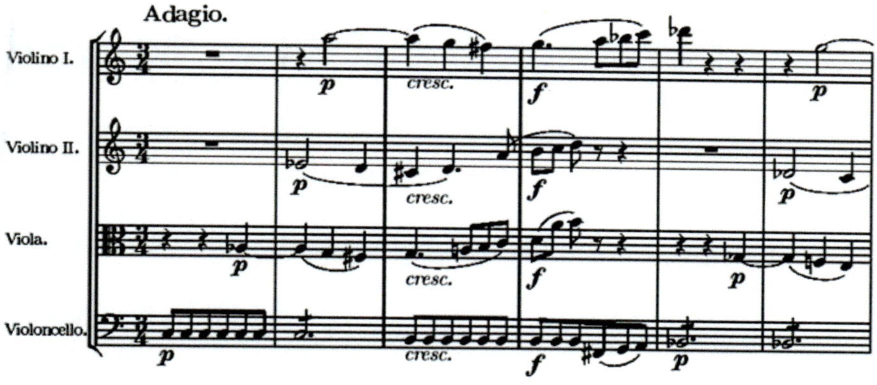

Fig. 3.4 The String Quartet No. 19 in C Major, K. 465 by Wolfgang Amadeus Mozart

achieve. In many cases, the powerful overall sound of the big orchestra comes to the fore, although soloists or the conductor may focus on the public's attention.

3.5.3 Pursuit and Escape in the Dynamics of the Fugue

In the animal world, pursuit and escape are basic patterns of locomotion, at least since the Cambrian revolution, called the "biological Big Bang" (approximately 543 million years BP). Young animals prepare for pursuit activities, and children play the game *mouse and cat* or *cops and robbers,* even in modern societies where predation is not a shared experience. In Western movies, endless pursuits fill the scenario (mostly on horses); in the James Bond series of movies, different kinds of this play are shown: in cars or other mobiles, in boats, planes, helicopters, or running on roofs (in Siena: cf. Wildgen [46] for an analysis of *movie-physics* in James Bond).

The name fugue derives from Latin *fuga,* which means escape. In the act of musical pursuit and escape, the first movement introduces a theme (by one voice, e.g., the tenor); it is picked up by another voice, which, in many cases, enlarges the subject and then passes it on to yet another voice. The voices already introduced continue on their course; they virtually "chase" the first, i.e., the "stolen" theme. A simple form of this game is the popular canon; while one part of the choir continues to sing, the other part starts with the melody, and so on, if more than two parties are involved. Eventually, all parts of the chorus close on a final chord. The individual parts of the choir go through the song in a time-delayed manner, creating a polyphonic structure. In J. S. Bach's work on the fugue, two-fold and three-fold canons are represented, and a kind of mirror theme may be added to the respective melody. He composed 14 elaborated fugues over the first eight notes of an aria from the Goldberg Variations. Bach's "The Art of the Fugue" is an artistic presentation of the astonishing richness

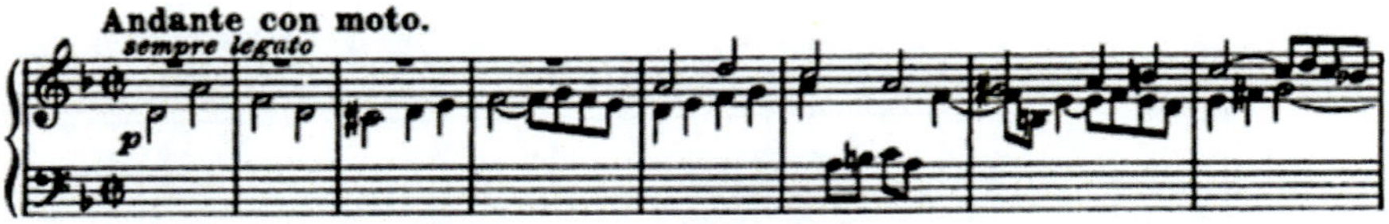

Fig. 3.5 Topic (Dux) and response (Comes) in Bach's Contrapunctus I (eight bars shown)

of possible variations. In the following, a sketch of the dynamics of this pursuit and escape, based on the musical form of the fugue, is given.

It is characteristic of the fugue that the theme repeatedly occurs in different voices, variations, and keys. The Comes is a way of answering a topic, often on the fifth step of the key.[20] The individual fugues of the work contain more complex forms, double or multiple fugues, reflections, and symmetries, and have an exemplary, almost didactic character, while, at the same time, the variety of possible forms and thus the freedom of the composer become clear (Fig. 3.5).

The Contrapunctus I is written in D minor; accordingly, the theme begins with D and ends with D in the fifth bar. Simultaneously with the end of the theme (Dux in the alto voice), the Comes in the soprano voice begins, shifted by a fifth, with A and ends with A at the beginning of the ninth bar (not shown in Fig. 4.7); in the bass, the theme is taken up again in bar 9 and the Comes in the tenor part in bar 13. A few episodes follow. At the end (coda) of the fugue, the tenor part resumes the Comes in A (cf. the schematic analysis in Heinrich [17]: 103). The C sharp in bar 3 can be considered the symmetrical midpoint of the dux,the eighth notes in measure 4 form a "movement impulse" since they repeatedly serve as "motivic initiator of the opposing voices, countersubjects and interludes" (Eggebrecht [9]: 54).

The musical effect of the fugue consists, on the one hand, of the strong coordination of the voices through polyphony, and, on the other hand, of the counterpoint, which introduces different kinds of countermovements and melodic conflicts. The development of the fugue or, generally speaking, of a more extensive orchestral work can be described as the unfolding of a musical score. On the one hand, there are modulations, i.e., a given structure is transposed and varied. Leyton [24] describes transposition as a "telescoping" similar to the optical telescope; the rings are shifted so that the same image appears at different distances. In music, this does not happen continuously but along a hierarchy (e.g., by a shift to the fifth). The unfolding proceeds from a regular expectation hierarchy, a standard, and an ideal image. If this expectation is broken, it creates an aesthetic appeal (ibid. 151ff.). The essential momentum consists of transfer effects from (past) given, known, to (future) new musical structures.

The instrumental music of Bach (and the baroque music that ends with Bach) has a playful character (see the ancestral emotion PLAY in Sect. 3.3.2) in several respects:

---

[20] "Topic theory" was developed by Leonard Ratner (1916–2011); cf. Mirka [29].

- There is a finite number of elements and forms defined by scale, keys and tonal steps, classes of topics (Dux, Comes), and voices: bass to soprano; canon or two-voices, four voices.
- There are several operations: repetition (with variation), inversion, mirroring, and crab motion.
- The procedural structure is defined, at least in outline, by an exposition at the beginning, a sequence of dux-comes figures (with interludes), and a conclusion (coda) with a return to the original key.
- In Bach's fugue work, a macrostructure can also be observed in the sequence of the individual fugues (Contrapunctus I–XVIII as numbered by Heinrich [17]). At least the simple fugues (I–III) are followed by the double fugues and later the mirror fugues. The nature and reasons for these structures are, however, disputed.

The classic game chess shows a similar array of figures and moves. Possible macrostructures can be found in collections of famous games. The emotional structure is also part of the game to a certain extent. Thus, the game of chess is geared toward competition, struggle, and victory (checkmate) with intermediate steps such as winning chess pieces or opening or closing paths (possible operations). In music, the tempi, accelerations, loud–quiet transitions, changes from major to minor, and the dominance of voices/instruments are emotional triggers in the musical game.

The Art of Fugue belongs to Bach's late work. It contradicted the dominant zeitgeist, which preferred the elegant (French) style, easier access for the general public, and a reference to the word/song. In the Viennese Classicism period (approximately 1770–1830), Bach's work fitted into the new image of musical genius and became a model for many composers (e.g., for Beethoven).

The playful character of the music becomes even more evident in the second Viennese school, especially in the work of Schönberg and the serial music of the twentieth century. Hermann Hesse emphasized this aspect of music as an abstract game in his novel "The Glass Bead Game" (begun 1930; first edition 1943). In European music, the 12-tone scale, in which major and minor chords, as well as the tonal step model, had been eliminated, was a drastic change in the "rules of the game". If the guiding function of certain tones or tonal steps is canceled, the tonal architecture becomes freer and more open. To compensate, however, new rules had to be introduced. For example, one rule states that in a sequence of tones in the 12-tone row, all tones must be realized either in sequence or as chords, avoiding repetitions that could create a focus.

3.5.4 Quasi-narrative Structures in Music and Their Dynamic Equivalents

In many publications (especially in the tradition of structuralism), narrative structures of music are put to the fore (see Grabócz [12 and Tarasti 36]). However, in most cases, narrative patterns in music are language-independent because they exhibit more basic dynamic patterns known from the theory of dynamical systems. The processual, quasi- "narrative" character of music refers to a more general behavior not expressed specifically through language. The archetypal "narrative" beyond language has three levels of dynamics that complement each other:

1. General structures of events. They characterize the envelope of the entire movement. In terms of the theory of dynamic systems (cf. Wildgen [42], second part), there are several basic types of dynamics: The simplest development of an unstable dynamic system in time is the "fold", characteristic of the beginning and the end. More complicated courses of events are the bipolar change between states A and B, the change with three states A, B, and C, where B represents a metastable transition zone, and the continuation of this series. These dynamic patterns can be called "narrative" because they share dynamic characteristics with storytelling.
2. The second level concerns the specific types of locomotion, their patterns of movement, and the coordination of several components of movement (such as legs and arms) or different persons in joint movements. Music develops this basic set of motion patterns into a rich field of modulated motions experienced acoustically.
3. Moving toward a goal often happens in phases. In a mountain hike from one mountain hut to another, the huts are the resting points; in between, there are steeply rising, descending, flat, or smaller up-and-down sections. There may be obstacles (chasms, ravines, rock walls, rivers, etc.) that must be overcome. There are signs of fatigue (at the end of the hike) and brisk stride (in the morning). Similar profiles are conceivable in larger pieces of music, e.g., in overtures to an opera or symphonies.

A plausible hypothesis says that musical dynamics are closely linked to movement dynamics and reflect them. The dynamics of movement can be divided into phases. There is a beginning and an end (this applies on a small scale to subunits of a piece of music, the movements of a sonata, and the stanzas of a song). There are different tempi and rhythms, acceleration and deceleration, and there are climaxes and steps to the climax or down to the endpoint. There are tensions between the individual forms of movement, for instance, height and energy differences. Music may be intentional in some of its forms. However, it is self-referential, like dance or children's games.

3.5.5 *Muso-mathematics and Computerized Music*

Mathematical considerations were already central to musical thinking and composing in antiquity. We cannot review the further evolution of the dialogue between music and mathematics. Still, in the twentieth century, with synthesizers and computer-aided music, mathematically sophisticated types of music became dominant, at least in the avant-garde scene. Muso-mathematical considerations played a role in musical composition since Claude Debussy (1862–1918) and Olivier Messiaen (1908–1992). The topological network of tunes already explored by Leonard Euler in 1739 was rediscovered at the end of the nineteenth century and gave rise to the "Tonnetz"-architectures of music based on mathematical group theory. The "Tonnetz" (German expression for a musical network) is a torus of tunes based on the intervals: fifth, major, and minor third. It can be flattened for easier understanding and use; cf. the description in Andreatta ([2]: 81–84) and Bergstrom et al. [3].

Practical consequences of the Tonnetz-intuition for composition can already be found in the work of Claude Debussy. "Many of his works are characterized by complex symmetries, transpositions, and proportions" (Pagano [30]: 90). Even geometrical proportions like the Golden Section, already used in antiquity and specifically in Renaissance art (e.g., by Leonardo da Vinci), play a fundamental role (cf. ibid.). Thus, music is systematically related to spatial patterns transposed into the temporal mode that dominates musical performance. Messiaen not only considered prime numbers and palindromes in his compositions, but he introduced non-Western scales and free rhythms (ibid. 92f.). In his "Traité de rythme, de couleur et d'ornithologie", published 1949–1992 (see Messiaen [28]), he gives a table of the prime numbers smaller than the number 100. He discusses their role in musical composition (cf. Papadopoulos [32]: 541). Papadopoulos made a mathematical analysis of the rhythms Messiaen chooses from historical (Greek and Indian) sources and his innovative use of mathematical insights in his compositions (ibid. 540–564). Moreover, Messiaen had an intuitive notion of color in music, a suggestion that other musicians of the twentieth and twenty-first centuries followed. If the musical style, its use, and the transformations it allows refer to geometrical notions as those explained by Felix Klein's Erlangen Program (i.e., basically group theory), the next generations of composers took up the suggestions of Thom's morphogenesis (cf. Andreatta [2]) and the concepts introduced by Chaos Theory and Fractal Geometry, for instance, Georgy Ligeti (1923–2006). Pagano [30] analyzes Ligeti's piano studies: Vertigo, L'Escalier du diable, and Columnia infinita (written around 1991). She writes: "the Devil's Staircase is a mathematical function and the piece effectively describes it" (ibid. 96).

These hints may be sufficient to show that mathematical models of music are not only a possible means to understand music; musicians also use them to compose music and open new fields of musical performance and understanding. In many treatises, music is understood as a kind of language, and the analysis of language is used as a foil to the analysis of music. The following section will examine the analogy between music and language and the differences in their morphogenesis.

3.6 Similarities and Differences in the Morphogenesis of Music and Language

Rousseau believed that the origin of language lies in music and the expression of passion through music. This origin, in his view, is obscured, blurred, and even erased in language by the deliberate reduction to a small set of sound shapes. In poetic language, a mood or temper remains that refers to passion, called "accent" by Rousseau. In ordinary discourse, "articulation" dominates. This historical answer to the question precedes the development of modern evolutionary theory and considers language as a degeneration of music. The question of the evolutionary relationships between the morphogenesis of language and music is still unsolved. The most plausible answer assumes a kind of co-evolution.

3.6.1 The Co-evolution of Language and Music as a Source of Common Features

Language and music probably have, as their sources, specific types of bodily motion and locomotion, their rhythm and resilience, the coordination of moving body parts—legs, arms, hands, head, and eyes—as well as the coordination in a human group activity such as hunting and making/handling of tools. If sounds served as signals in the evolution of language, as shouts or drums did, later on, speaking in songs could become music.[21] The transitions are fluid. In many cultures, ritual utterances have a rhythmic, even melodic, diction, usually with a restricted tonal range. There are two significant patterns separating language and music:

(1) In early and exotic cultures and Western mass culture, the songs are characterized by constant, almost monotonous repetitions. This moment is probably "inherited" from the signal function of animal sounds. The phenomenon of repetition can appear in language as echolalia and ritually monotonous speech, for instance, in the litany. This phenomenon is, however, marginal in language. In classical musical art, repetitions are common. In complex musical forms, they are masked by variation or counterpoint and later by serial effects.
(2) The rhythm of music goes beyond the simple patterns of intonation (i.e., variation in pitch) of language. Especially in cultures without pronounced polyphony, complex polyrhythmic structures arise. This may be related to the function of

[21] Pygmy women in Central Africa behave differently than men: "Women's speech is more song-like than men's, and they accompany each other's utterances with sung expletives that contribute to increasing the volume and distinctive melodiousness of their conversations." Lewis ([25]: 83). As pygmy populations may be characteristic of early hunter–gatherer populations, this feature could point to a source of the morphogenesis of music in female conversation. However, as we presume that music has the same age as language, both origins are very far from actual music and language use. Therefore, such a hypothesis cannot be validated. Cf. Sect. 6.1 for the origin of language.

drums as signaling instruments and the unfolding of the possibilities of this acoustic source.

The emergence of musical scales, the normalization of intervals, is a later process. The singing voice is often variable and glides in the intonation without the attunement to the norms of a specific music culture. Early simple instruments were not exactly tuned or consonant. In addition to the natural disposition of the ear, which is sharpened in the context of instruments, instrumental music may have been the main promoter of development. Stumpf ([34]: 40) compares their effect with writing. For example, the production of flutes and pipes sets standards for sound intervals. A particular tone scale becomes the social norm by using the same instruments over generations or their exact reproduction. This development also fixes the intervals of the singing voice, i.e., it counteracts the natural uncertainty of the singing voice. Similarly, writing standardizes first the texts, for example, in the hieroglyphic texts of the Egyptian temples and tombs, then the spoken language; cf. the development of an alphabetic writing system that fixes its sound structure. With the standard scriptural norms, such as in classical Latin or Arabic (and other *frozen* (dead) standard languages), the grammar and style of a language are forced into the harness of strict rules. Something similar happens with music, first in the ritual context (for example, in the religious festivals of the Greeks and Gregorian chant), then in secular music.

The divergence of language and music in later periods led to conflicts between musical and linguistic norm structures. As long as the religious (and, by analogy, the secular) song was monophonic and limited in length, text and music could run in parallel and repeat or continue simple patterns. However, as the lyrics became longer and less repetitive, an extension of the musical pattern became necessary. As a result, the differences between linguistic and musical norms sharpened. The music of a song can obscure the linguistic message or even reduce it to a melodic line. The increasing dominance of musical accompaniment (in the baroque, for example, through the organ) competed with the message and text character of the song. In opera (for example, in Wagner) or lyrical, musical performances (such as Henze or Schönberg), the listener must know or read the text to understand the lyrical intention.

The zones of coordination between language and music are "hinge joints" of multimodal communication,[22] i.e., although, in general, these symbolic forms have their character, a common denominator exists, such as regularities of adjacency or subordination. There are (at least) two major coordination axes of multimodal symbolic forms:

(1) *The lexicon of topoi*

Known from rhetoric, topoi (sometimes called "commonplaces") are subject areas that can be the topic of speech and discourse. They crystallize as relevant in historical development and constitute a finite inventory. Agawu ([1]: 41 ff.)

[22] In the human skeleton, hinge joints like a knuckle or elbow allow the coordinated movement of limbs. Similarly, different modes of symbolic communication, e.g., language, music, and dance, have specific zones in their organization where their interaction is crucial, for instance, rhythm in music and dance or meter and rhythm in lyrics.

compiled lists of topoi for particular periods of Western music. In the Baroque period, a narrow, word-oriented meaning assignment of musical figures and a coarser global content assignment (res) were distinguished. At a time when instrumental music competed with and surpassed the vocal, the paradigmatic change from textual meaning to pure musical meaning became obvious. Only in the program music of the nineteenth century did tendencies toward "speaking" music reappear.

(2) *The punctuation* (also called periodicity; see Agawu [1]: 75 f.).

It determines the beginning, climax, and end of a musical phrase. The ending is called "cadence". This connects the listener's expectations, surprise, suspense, satisfaction, and tranquility. Overall, processual phenomena dominate the music. These aspects of music can be compared with discourse structures in language and dynamics in nature that are partly predictable and partly surprising.

Language and music have similar spatial and temporal archetypes[23] and are also similar in regions of contact and coordination within multimodal forms. Nevertheless, they differ fundamentally in the semiotic content, the material, and the creation and organization of meanings.

3.6.2 Semantic Constructions in Language and Music (Valences and Narrative Patterns)

From the perspective of (linguistic) structuralism, the basic difference between music and language is that music is organized on a single layer. In this view, musical signs convey meaning without using a preparatory, basic level as phonology in language. Because musical meanings are not denotative, discrete, and organized in a lexicon, it does not make sense to project the grammatical levels (phoneme, morpheme, word, and sentence) of language into the realm of musical signs. However, the comparison of the construction principles could be promising, although highly speculative. In Western tonality (since the Renaissance), one can identify examples of similar constructive principles in music and language. The musical model of tonal steps can be compared in its architecture with the valence steps in language. The correlate of the musical tonic (the basic tonal level) would be the nominative, which is often unmarked and stands in the intransitive sentence (valence 1); the dominant step in music or second most important (and most frequent) level correlates with the accusative case that is added in the transitive sentence (valence 2); the subdominant, or the third most important step in tonal music corresponds to the dative case.[24] The dependency or dominance relations in music have an anchor in the natural musical scale, which is physically (and physiologically) based on the overtone series (cf.

[23] "Archetype", from Greek ἄρχω "to begin", means a form or behavior that existed already at the beginning of morphogenesis and remains relevant for further developments.

[24] Cf. For the classification and dynamics of case systems in natural languages Sects. 6.4 and 6.5.

this chapter). In the case of language, valence has its natural foundations in event and action archetypes or basic scenarios such as locomotion (valence 1), action on an object/subject (valence 2), and a transaction in giving/taking (valence 3). The ontological basis is different; this corresponds to the different functions and modes of action of the two symbolic forms; however, the gradations in the basic patterns are comparable. This anchoring of symbolic systems in principles of nature (the phenomenal environment) can be described as archetypal (see Wildgen [41, 42]). Beyond these archetypes, a broad field of contingent, context-dependent, and seemingly random regularities exists. They form the arbitrariness that Saussure has placed at the center of his theory of language.

The theory of music following the semiotics of A. Greimas sees in the narrative patterns the common denominator of language and music (cf. Tarasti [36]: 24–28 and Grabócz [12]). However, the analysis of thematic, narrative structures in music relies heavily on comparing music with texts and the effect written texts have on the composer of a piece of music. First, the music composer may have written both operas and instrumental music so that one can infer the narrative content of similar instrumental passages from the narrative contents of the opera,second, literary models (dramas, novels, and poems) may have served the composer as a guideline. The narrative analysis of music depends on a historically given bond between literature and music. The narrative music analysis is pointless if the bond is weak or inexistent. Language and music can be traced back to more general patterns rooted in the semiotic ontology common to both symbolic forms. Cf. Chap. 7 for the analysis of a basic ontology of symbolic forms and semiosis.

The temporal organization of music arises from the "refraction" (Schenker), i.e., the temporal unfolding of the static chord. The realization of latent harmonies in time creates a kind of "sound space–time" in which musicians and listeners move. In polyphonic music, different refractions appear side by side (ordered by the rules of counterpoint). The composition (in music) and the syntactical formation of sentences and texts in the (respective) language not only inherit the basic similarity of music and language but also incorporate specific conventions and inventions so that, at the surface, the correlation of music and speech is obscured. Nevertheless, in the creative design of music, i.e., in the composition of new musical works and of language art (in literary innovation), the fundamental structures and creative forces become again visible (cf. Wildgen [47]: part II for examples).

3.7 Musical Meaning and Context

Musical meaning depends on the before and after of the musical event. However, this integration into the temporal events works differently than in language. Meaning in music is not built up in hierarchical and schematic constructions of elements, as is assumed for the morphology and syntax of a language. It is rather a kind of overlay, the addition of effects, the integration of local phenomena into larger wholes, and the emergence of a global meaning from many interacting forces.

In addition to this narrow context, there is another concept of context related to discourse. In Foucault's sense, discourse is a complex ("dispositif") of various functions, usually a heterogeneous structure that has emerged from various social conditions and objectives and often reflects power relations. In music, the political component has received great weight, especially in the context of the French Revolution. It "was about integrating music into the overall national concept" (Harnoncourt [16]: 27). The "Conservatoire" undertook the task of coordinating national music education. For this purpose, "everything that speaks—requires understanding—should be eliminated from music" (ibid.). This reorientation and the newly created institutions are still effective today and have partially erased the earlier traditions of musical education and thus caused a culture break in music history (ibid. 27f.).

After the failed revolution in Germany in 1848, "romantic hopes faded," and.

> the realm of the aesthetic increasingly came to represent a retreat from life (arguably still a kind of political gesture). This moment coincided with Hanslick's formalist agenda, and was followed by the Wagnerphil French symbolists' search for hermetic poetry that would emulate music's apparent disconnection from the everyday world of living (Clark [7]: 177).

In the twentieth century, political measures were also against musical styles, such as the early rejection of jazz (first in the USA, later in Europe), rock and pop music in the Eastern Bloc, and atonal music in the conservative European circles. In North Korea, Western music forms are banned, and the Taliban in Afghanistan do not want to tolerate such forms. Consequently, there must be a relationship between musical discourse and political/religious power. Music transports values that can promote a revolutionary change and transition (in the case of the French Revolution) or enter into conflict with the norms of a rigid, authoritarian, or religious rule. On the other hand, music may also be an escape from the insecurities of social and political developments and thus reduce the impact of historical contexts.

References

1. Agawu, K.: Music as Discourse: Semiotic Adventures in Romantic Music. Oxford U.P., Oxford (2009)
2. Andreatta, M.: Stabilité structurelle et morphogénèse musicales : la théorie des catastrophes et la recherche musicale. In: Papadopoulos, A. (ed). René Thom : Portait mathématique et philosophique, 381–396. CNRS Éditions, Paris (2018)
3. Bergstrom, T., Karahalios, K., Hart, J. C.: Isochords: visualizing structure in music. In: Proceedings of the Graphics Interface 2007 Conference, May 28–30, Montreal (Canada) (2007) https://www.researchgate.net/publication/221474662_Isochords_visualizing_structure_in_music
4. Berendt, J.E. : Das Jazzbuch. Von New Orleans bis Free Jazz. Fischer, Frankfurt/Main (1970)
5. Bogue, R.: Deleuze on Music, Painting, and the Arts. Routledge, New York (2003)
6. Chella, A.: A cognitive architecture for music perception: exploiting conceptual spaces. In: Zenker, F., Gärdenfors, P. (eds.) Applications of Conceptual Spaces, Synthese Library, vol. 359, pp. 187–203. Springer, London (2015)
7. Clark, D.: Musical autonomy revisited. In: Clayton, M., Herbert, T., Middleton, R. (eds.): The Cultural Study of Music: A Critical Introduction, 2nd edn., pp. 172–183. Routledge, New York (2012)

8. Eicken, A.: Der Affektbegriff in der Musik des Barock. https://silo.tips/download/der-affektbegriff-in-der-musik-des-barock (2000)
9. Eggebrecht, H.-H.: Bachs Kunst der Fuge. Ercheinng und Detung Piper, München (1985)
10. Finnigan, R.: Music, experience, and anthropology of emotion. In: Clayton, M., Herbert, T., Middleton, R. (eds.) The Cultural Study of Music, pp. 352–363. Routledge, New York (2012)
11. Gibson, J.J.: The Ecological Approach to Visual Perception. Houghton Mifflin, Boston (1979)
12. Grabócz, M.: Musique, narrativité, signification. L'Harmattan, Paris (2009)
13. Habib, M.B.: Fossils reveal when animals started making noise: for billions of years earth was quiet. Then Life got Loud. Sci. Am. **326**(1), 42–47 (2022) https://www.scientificamerican.com/article/fossils-reveal-when-animals-started-making-noise/
14. Haken, H.: Principles of brain functioning. A Synergetic Approach to Brain Activity, Behavior, and Cognition. Springer, Berlin (1996)
15. Halle, J.: From Linguistics to Musicology. Notes on Structuralism, Musical Generativism, Cognitive Science, and Philosophy. In: Brandt, P. Aa., do Carmo, R. Jr. (eds.): Sémiotique de la Musique / Music and Meaning, pp. 287–311. Presses Universitaires de Liège, Liège (2015)
16. Harnoncourt, N.: Musik als Klangrede, 4th edn. Wege zu einem neuen Musikverständnis. Residenzverlag, Würzburg (1982)
17. Heinrich, A.: Bach's Die Kunst der Fuge. In: A Living Compendium of Fugal Procedures with a Motivic Analysis of All the Fugues. University Press of America, Washington D.C. (1983)
18. Hiekel, J. P.: Von der Nötigung zu einer eigenen Sprache und vom „Sinnverlust" als Gewinn. Anhand des Themas „Klang und Bedeutung" ergibt sich eine neue Sicht auf die Musikgeschichte der vergangenen sechzig Jahre. Neue Zeitschrift für Musik 1_2014, 90–99 (2014)
19. Kelso, J.A.S. (ed.): Learning To Live Together: Promoting Social Harmony. Springer, Cham (2019)
20. Kühl, O.: Musical Semantics. Peter Lang, Bern (2007)
21. Larson, S.: Musical Forces: Motion, Metaphor, and Meaning in Music. Indiana University Press, Bloomington (2012)
22. Leman, M.: Embodied Music Cognition and Mediation Technology. MIT-Press, Cambridge (Mass) (2008)
23. Leyton, M.: A Generative Theory of Space. Springer, Berlin (2001)
24. Leyton, M.: Musical works and maximal memory stores. In: Mazzola, G., Noll, Th., Lluis-Puebla, E. (eds.) Perspectives in Mathematical and Computational Music Theory, pp.116–152. epOs-Music, Osnabrück (2004) https://citeseerx.ist.psu.edu/document?repid=rep1&type=pdf&doi=7666b3e543906df933569bbce5edde8c128dbb26
25. Lewis, J.: BaYaka Pygmy Multimodal and mimetic communication traditions. In: Dor, D., Knight, Ch., Lewis J. (eds.): The social Origins of Language, pp. 77–91. Oxford U.P., Oxford (2014)
26. Luhmann, N.: Liebe als Passion. Zur Kodierung von Intimität. Suhrkamp, Frankfurt/Main (1982)
27. Mazzola, G.: Geometrie der Töne. Elemente der mathematischen Musiktheorie. Birkhäuser Verlag, Basel (1990)
28. Messiaen, O.: Traité de rythme, de couleur, et d'ornithologie (1949–1992); seven volumes. Reprint Leduc, Paris. First volume (2004) English translation of the first volume by Baggech, M., A.. Graduate Faculty of the School of Music, Norman, Oklahoma (1998) 9914415.PDF
29. Mirka, D. (ed.): The Oxford Handbook of Topic Theory. Oxford U.P, Oxford (2014)
30. Pagano, S.: Mathematics and labyrinths in twentieth century's music. In: Emmer, M., Abate, M. (eds.): Imagine Math 6. Between Culture and Mathematics, pp. 89–99. Springer, Cham (2018)
31. Panksepp, J., Biven, L.: The Archaeology of Mind: Neuroevolutionary Origins of Human Emotion. Norton, New York (2012)
32. Papadopoulos, A.: Mathematics and group theory in music. In: Lizhen J., Papadopoulos, A., Shing-Tung Yau (eds.) Handbook of Group Actions, vol. 2, pp. 525–572. Higher Education and International Press, Bejing-Boston (2015)

33. Parncutt, R.: Harmony: The Psychoacustical Approach. Springer, Berlin (1989)
34. Stumpf, C.: Die Anfänge der Musik, 1911. Reprint. Georg Olms Verlag, Hildesheim (1979)
35. Sundberg, J.: The Science of Musical Sounds. Academic Press, San Diego (1991)
36. Tarasti, E.: La musique et les signes. Précis de sémiotique musicale. L'Harmattan, Paris (2006)
37. Thom, R.: Stabilité structurelle et morphogenèse, Interéditions, Paris (English translation: Structural stability and morphogenesis. Benjamin, Reading, 1975) (1972)
38. Thom, R.: Esquisse d'une sémiophysique: physique aristotélienne et théorie des catastrophes. Interéditions, Paris [English translation: Semiophysics. A Sketch. Addison-Wesley, Boston 1990] (1988)
39. van Leeuwen, T.: Speech, Music. Sound. Red Globe Press, London (1999)
40. Wermke, K., Mende, W.: From emotion to notion: the importance of melody. In: Decety, J., Cacioppo, J.T. (eds.) The Oxford Handbook of Social Neuroscience, pp. 624–648. Oxford U.P, New York (2011)
41. Wildgen, W.: Catastrophe Theoretic Semantics: An Elaboration and Application of René Thom's Theory. Benjamins, Amsterdam (1982)
42. Wildgen, W.: Process, Image, and Meaning: A Realistic Model of the Meanings of Sentences and Narrative Text. Benjamins, Amsterdam (1994)
43. Wildgen, W.: The Evolution of Human Languages: Scenarios, Principles, and Cultural Dynamics. Benjamins, Amsterdam (2004)
44. Wildgen, W.: Kognitive Grammatik. Klassische Paradigmen und neue Perspektiven. de Gruyter, Berlin (2008)
45. Wildgen, W.: Visuelle Semiotik. Die Entfaltung des Sichtbaren. Vom Höhlenbild zur modernen Stadt. {Visual Semiotics. The Unfolding of the Visible. From Cave Pictures to modern Towns] Transcript, Bielefeld (2013)
46. Wildgen, W.: "Movie physics" or dynamic patterns as the skeleton of movies. In: Wildfeuer, J., Bateman, J. A. (eds.): Film Text Analysis: New Perspectives on the Analysis of Filmic Meaning (Routledge Advances in Film Studies), pp. 66–93. Routledge, New York/London (2016)
47. Wildgen, W.: Musiksemiotik: Musikalische Zeichen, Kognition und Sprache [Semiotics of Music. Musical Signs, Cognition, and Language]. Königshausen & Neumann, Würzburg (2018)
48. Wildgen, W.: Natural sources of musical meaning: the ecology of perception, motion, and archetypal emotions. In: Contribution to the 15th International Congress on Musical Signification, in Barcelona, 15–19 June 2022

Chapter 4
The Morphogenesis of Visual Symbolic Forms (Art, Architecture, and Urban Structures)

Abstract This chapter begins with a review of visual analysis methods since the twentieth century and the relevance of morphogenesis to it. The evolution of visual art points to morphogenetic principles already visible in Paleolithic "art". Next, the morphogenetic analysis focuses on innovations in art since the Renaissance (with a focus on Leonardo a Vinci), the late modernity (William Turner, Henry Moore), and the twentieth century (Pollock and Joseph Beuys) with side-remarks on auto- and meta-reference in paintings by Pablo Picasso and Salvador Dali. In the remaining sections, visual artifacts beyond painting and sculpture, i.e., in architecture and urban design, are investigated regarding morphogenesis.

4.1 The Methods and Principles of a Morphogenetic Analysis of Visual Forms

In overview, there are different levels of access to visible shapes and visual art:

- The *micro-level* (physiological and neural processes). It is the precondition for visual perception, imagination, and the corresponding memory. In the psychology of perception dealing with humans (and higher animals), tests and experiments on the behavior of human (and animal) subjects led to relevant insights. The work of Arnheim may stand for this tradition in the analysis of art (see [1] and the next section).
- The *phenomenological* level. On the one hand, visual communication builds on differences (values) in visual perception. On the other hand, it draws on historically created and changing conventions (rules of visual understanding). The comparison of recurring motifs in art history is a significant access to the analysis of change and stability in the morphogenesis of art. The methods of iconography and iconology are a response to this demand. We shall discuss these theoretical enterprises in a second section.

W. Wildgen, *Morphogenesis of Symbolic Forms: Meaning in Music, Art, Religion, and Language*, Lecture Notes in Morphogenesis,
https://doi.org/10.1007/978-3-031-25651-6_4

4.1.1 Gestalt Principles of Visual Art After Arnheim

The basic principles of gestalt psychology, developed particularly in the experimentally oriented Berlin school (Wertheimer, Köhler, Koffka), were the basis for Rudolf Arnheim's psychological work on visual art.[1] His book: *Art and Visual Perception: A Psychology of the Creative Eye* (1954) made him famous. As the title suggests, seeing is an active process in which the eye/brain generates patterns based on light stimuli. These creatively developed shapes result from a morphogenetic process. It involves the organs of sight and reacts to the affordances of the visual environment. The active involvement of the cognitive system becomes particularly clear in the case of hallucinations, the completion of incomplete forms (Kanizsa contours), and the attribution of causal relationships. Two principles emphasized by Rudolf Arnheim will be applied in the analysis of Leonardo's "The Last Supper":

(a) The Principle of Centrality. Every scene and every image is broken down into a center of visual attention and an associated periphery. In the case of framed pictures, this center depends on the frame's shape (rectangular, quadratic, or round). In addition, sub-centers that are visited longer or more frequently by eye saccades can be detected, and transition zones between different sub-centers can be marked. This way, the visual field is structured depending on a specific viewing time. Various directions away from the center (like up-down, left–right) can also be highlighted and contribute to the meaning of paintings.
(b) The Force Fields. Rudolf Arnheim, like James Gibson, emphasizes the dynamic character of visual perception. Thus, a static rectangle that is white (empty) can be understood as a force field determined by the diagonals on the one hand and the center vectors on the other. The latter is related to the above principle; the former results in movement to the corners and left–right asymmetry if one of the diagonals dominates in the image. Overall, the image field is formally a visual vector field into which the specific movements of the visual attention are inserted and adapted depending on the shapes or colors.

Overall, there is a layered level structure of visual perception. First, there are the physical stimuli such as contours, gradients of brightness, colors, and surfaces; secondly, gestalt processes in perception take place that constructively change, supplement, and fill out these stimuli. Thirdly, there is the attribution of meaning, which builds on these perceptual results. Finally, existing memory patterns or biographical (indirectly also historical) experiences can be modified and inserted into a larger context. The Berlin School of Gestalt Psychology, Gibson's research, and modern neuro-vision research have shown that the first two levels can be assessed

[1] He studied psychology (regarding art) in Berlin and received his doctorate in 1928 with a thesis on the expressive behavior of human faces and hand movements. He was interested in film from an early age (in Rome until 1939). After immigrating to the USA, he was first appointed by the Guggenheim Foundation commissioned to research the contribution of Gestalt psychology to art science. He then taught art psychology at various American universities until old age. Cf. for an overview Verstegen [32].

using experimental methods. The memory component is the proper realm of symbolic forms, although it already plays a role on the second level.

4.1.2 *Visual Art as Symbolic Form; The Impact of Iconography and Iconology*

In Hamburg, Cassirer had good contacts with the Warburg circle and Aby Warburg's cultural studies. In various writings, especially towards the end of his life, he regards art as a particular symbolic form. It is a language of intuitive symbols without the abstract categorization typical of natural languages and its segmental-sequential organization (cf. Cassirer [7]: 152f.). The art-historical dimension of Cassirer's philosophy was expanded in particular by Erwin Panofsky and Fritz Saxl. Panofsky distinguished three levels of meaning in the visual arts (cf. Panofsky [26]: 32ff). The first two relate to the subject (for instance, in a picture). A distinction is made between the primary or natural subject and the secondary or conventional subject. The former concerns lines and forms and recognizable natural objects such as houses, animals, relationships, movement forms, and properties. The second type of subject identifies people or things based on conventionally assigned characteristics. For example, in a picture, there is a male figure with a knife (primary subject),this configuration refers to St. Bartholomew (secondary subject). To be able to identify the secondary subject, the viewers must be familiar not only with the pictorial tradition but also with literary traditions. Only then can they "read" the secondary meaning. The third type of meaning is the "actual meaning or content" (ibid. 40). In it, an author condenses the principles or perspectives of his epoch, society, and social environment. "The discovery and interpretation of these 'symbolic' values (…) is the subject of what we, in contrast to 'iconography', can call 'iconology'" (ibid. 41). One can say that the iconographic analysis is the empirically justifiable part of the analysis of a work of art. The iconological is the theoretical part that proposes possible connections and regularities of a more extensive range.

4.1.3 *Further Methods of Visual Analysis in a Morphogenetic Perspective*

Kress and van Leeuwen [15] introduce three interconnected systems of meaning creation in visual artifacts:

1. Information value. Different zones of the picture are associated with specific valences, e.g., left (less)–right (more) or up (desire, ideal)–below (real), central–peripheral (different information values in Asian culture) (cf. ibid. Chap. 6).
2. Salience. It is realized by qualities such as foreground versus background, relative size, a contrast in brightness or color, and differences in sharpness (ibid. 183).

3. Framing. Visible frames or dividing lines can connect or separate, indicating that certain contents are related or unrelated.

These criteria elaborate on basic principles already mentioned by Arnheim but specified in the perspective of social semiotics, i.e., the meanings contained in visual art and design due to social contexts and specific applications (for instance, in journals and advertisements).

Jean Petitot introduced the notion of non-genericity (or instability under variation) as a correlate of aesthetic quality. He speaks (Petitot [30]: 57) of significant and pertinent (spatial, temporal) structures marked by non-genericity. This morphological information serves as a basis for interpretative processes and thus constitutes a morphology of meaning ("morphologie du sens"; cf. Petitot [29], Brandt [2], and Bundgaard [4]).[2] Leyton [18] also addresses the dimension of the meaning of lines and curves in the context of his mathematical psychology. Several of these adventures into visual aesthetics will be used or referred to in the remaining sections of this chapter. In the morphogenetic analysis of visual art, the origins of this symbolic form must be the grounding floor.

4.2 The Origins of Visual Artifacts and Art

The beginning of graphic arts can be dated from the first appearance of concentrated color pigments around hominid sites. Barham (2001) reports that hematite (iron compounds often referred to as ocher) was recovered from a deposit near the Twin Rivers in Zambia. The pieces of hematite had been brought to the site, processed there, and rubbed on surfaces. One can conclude that these materials were used to color bodies or surfaces. Using such pigments points to a continuity that extends from the aforementioned archaeological sites (around 270,000 years BP) to contemporary hunter-gatherer cultures in the Kalahari. It is highly probable that the early Homo sapiens also used these (and other) colors to paint their bodies, objects, or wall surfaces. A larger body of paintings, line drawings, and sculptures did not appear until the later Stone Age (Palaeolithic), and the following discussion will focus on these products. Three types of figural symbolism ("art") can e distinguished: (1) Engravings on tools, (2) portable art (figurines) and sculptures, and (3) partially colored paintings in caves.

Ad 1: The first incised drawings on the stone were found in Africa and can be dated back to 70,000 years BP. One can distinguish between self-contained, mainly decorative forms, and iconic art, which refers to objects, animals, etc., through realistic contours and colors.

Ad 2: Palaeolithic sculptures. These sculptures are primarily small, like the famous "Venus statuettes" in many places in Europe and Western Asia. They generally

[2] Verstegen [33] sketched the consequences of the new paradigm, "dynamic semiotics", for visual analysis are. Caliandro [5] described some applications of the morphodynamic paradigm in the analysis of visual art.

emphasize the woman's sexual attributes. Other sculptures (e.g., of animals) are very realistic, such as the clay bison sculptures in the cave of Tuc d'Audoubet (cf. Leakey [16]: 174). The sculpture can even be a decoration on a weapon or ritual instrument.

Ad 3: Cave paintings occur primarily in an area north and west of the Pyrenees, e.g., in Périgord, Toulouse (France), and the Cantabrian provinces of Spain.[3] Together with the finds of sculptures in southern Germany and the Danube countries, they point to a cultural network south of the ice masses with a corridor between the northern European icy areas and the Alpine glaciers, i.e., from northern Spain to southern Russia. The herds of grazing animals that moved seasonally between the higher (colder) and lower (warmer) regions determined the lives of humans (and other carnivores who competed with them). One can compare the ecology of these cultures to reindeer cultures in northern Finland and northern Siberia today. The cultural contacts and the ecological basis of survival probably led to a shared (or at least similar) system of beliefs, myths, and rituals of which the cave paintings are an expression. The peak of cave painting extends from the end of the Aurignacian to the middle Magdalenian (around 16,000 years BP); this tradition was lost rather quickly towards the end of this period. In the decline of this culture, the images became smaller and were reduced to outlines and sketches.

The engravings (see 1) and paintings (see 3) were also simplified to symbols without iconic support. To those who used the cave, the meaning was known, but to those who did not take part in the ritual, they looked like actual ciphers from an unknown alphabet. Since we know that writing systems are more recent phenomena, we must interpret these symbols as memory-supporting notations. Even before a system of writing was introduced, and with it a corpus of knowledge held at the disposal of persons in prominent social positions, cultural knowledge has been a socially codified system of signs, anticipating later graphic modes of codification. After this step, cultural evolution could reach a level of organization that made writing possible and profitable. It only needed to be invented and further developed through its use.

Utility art and ritually integrated art form a continuum of forms. The so-called hand axes represent the first stage of development. It was perhaps preceded by a phase in which found objects (branches, bones, stones) were selected and only roughly prepared. With the first systematically manufactured stones (e.g., in Villefranchien, approx. 2 million years BP), the implementation of a traditional idea of form begins. It led to the serial production of objects (and prefigures technical innovations in the industrial revolution). Since hand axes were used both as weapons and for cutting and scraping, they partially replaced the function of teeth in other animals. In a way, hand axes are representations of canines (on the tip) or rows of teeth (on the tapered sides). These evolutionary aspects are further developed in Wildgen [35, 36].

[3] However, there are also sites in the southern Urals. Similar paintings in northern zones may have been chipped off by frost and disappeared into the rubble. Since some regions have no calcareous cavities, it is reasonable to admit that they may have existed in all Ice Age areas inhabited by modern humans. Recently cave paintings were discovered in Borneo. Therefore, it is plausible that they were a common feature of human culture in the migration inside and outside Africa.

The morphogenetic "germ" of Paleolithic art developed further in Egypt, Mesopotamia, the Minoan culture, and Greek antiquity. Finally, this high point of civilization was rediscovered and continued in the Renaissance. Therefore, we start with an analysis of the art and theory of art in Leonardo da Vinci; cf. for a more detailed analysis Wildgen [37], in French; [39], in English, and [41], in German.

4.3 Morphogenesis and Visual Art: From Leonardo da Vinci to William J. M. Turner and Henry Moore

At first sight, these two fields, morphogenesis and visual art, seem to have no common ground. Biological morphogenesis applies basic physical and physiological principles. One could say its realm is the physical world, "res extensa" in Descartes' terms. On the other hand, culture, media, and the arts are often considered abstract, if not spiritual. They are referred to as "res cogitans" in a Cartesian ontology. They seem to have no real-world substratum, to be non-accessible to scientific methods like measurement, experiment, statistical analysis, and mathematical modeling.

Contrary to Cartesian metaphysics and inspired by René Thom's epistemology, we assume an interaction between biological entities, e.g., human beings, and ecologies, mainly the living environment ("Lebenswelt"). Complex living systems have a kind of systemic telicity, i.e., a goal orientation. In its fight for survival, the organism creates stability and a projection into a possible future. To a certain extent, it can choose and channel possible evolutions, modify its life context, and thus avoid selective pressures. This telic dimension is the background of consciousness, which aims to control the interaction of humans with their environment. Rational strategies are the outcome of this trend. However, their impact is limited to lacunas of determinism left by more potent factors, which govern the overall evolution of the system. Art is also telic and applies principles of rationality (e.g., minimal means for maximal efficiency). However, its proper realm is intuitive creativity, which enables the appearance of new, astonishing shapes, constructions, and contents. In many instances, it overthrows established habits. It contradicts the rules of art academies (e.g., in the nineteenth century) and reacts to sudden political and societal shifts (e.g., after wars and economic crises). We shall present three case studies on Leonardo da Vinci, William J. M. Turner, and Henry Moore. They may stand for consecutive periods of morphogenesis in visual art: Renaissance (fifteenth and sixteenth century), the beginnings of impressionistic and expressionistic visual art (nineteenth century), and the twentieth-century shift to abstract forms. In a later section, avant-garde artists like Pollock and Beuys shall be dealt with separately.

4.3.1 *Morphogenetic Structures in the Art of Leonardo da Vinci*

In his "Trattato della pittura", Leonardo states that painters have as their primary aim the representation of two things: man and his mind ("l'uomo e la mente"; Pedretti [28]: § 180). The nature of man becomes visible and, therefore, accessible to the eye in the different "accidents" and movements and the proportions of his body parts (cf. Pedretti [28]: Terza Parte). To represent man and his mind, the artists must first create an appropriate pictorial space, the stage for the topic of the painting. Second, they must consider light and shadow on human bodies, the gestures of the hands, the postures of the head, and facial expressions. After that, they can distribute the topics of the painting on the surface. Finally, the landscape, sky, objects, animals, and persons accompanying the subject of the artwork must be arranged in space relative to light and shadow. Leonardo's art goes beyond the mimesis of nature; it manifests the meaning of space, light, and shadows in human experience. It displays the dramatic relations between persons and their embedding into the scenery (landscape, architecture, or the interior of houses). The beauty of a scene is the recovered meaning in the viewer's mind, the interpreter. Based on his experience, movement, and action, he must extrapolate from a static picture to a process that has caused the action and will bring it to a proper end later. In bodily movement, the dynamics of one body part have a counterpoise in another body part. The terms *balance, weight,* and *counterpoise* have the following meaning in the context of painting: A single body is in balance if the weight of the movement of one part of the body, e.g., the head, has a counterpoise in another, e.g., in the direction of the shoulders or the trunk. The balance could be easily realized if all bodies or body parts were static. However, this would make them appear "wooden", i.e., unanimated. The painter who wants to show the mind of the persons in the scene must show them in a *dynamic* balance. This objective means that, for Leonardo, natural dynamics is the source of bodily dynamics and, finally, the dynamics of the mind. Therefore, it has to start from the physics of objects, and bodies in space/time, resulting in the morphogenesis of meaning in a symbolic space. This morphogenesis differs depending on the kind of experience based on the human senses and their cognitive processing. Therefore, it is different for musical, visual, mythical meanings, and others. In the following, two paintings of Leonardo are analyzed: "The Lord's Supper" and "St. Anne with the virgin and the child" (for further details, cf. Wildgen [41], in German and 2004b, in French, where the pictorial traditions before and after Leonardo and his sketches in the preparation of the paintings are included in the analysis).

The topic of "The Lord's Supper" has been traditional in Christian art since the second millennium (cf. Wildgen [36]). The spatial framework of the painting is defined in terms of linear perspective. It creates the illusion of a deep hall with a rectangular tapestry on the sides and three openings going into a landscape in the background. The linear perspective has its origin in the head of Christ. The central point, Jesus' ear, is significantly related to the narrative content. Jesus has just uttered the prophecy that one of his disciples will betray him, and at the moment captured by

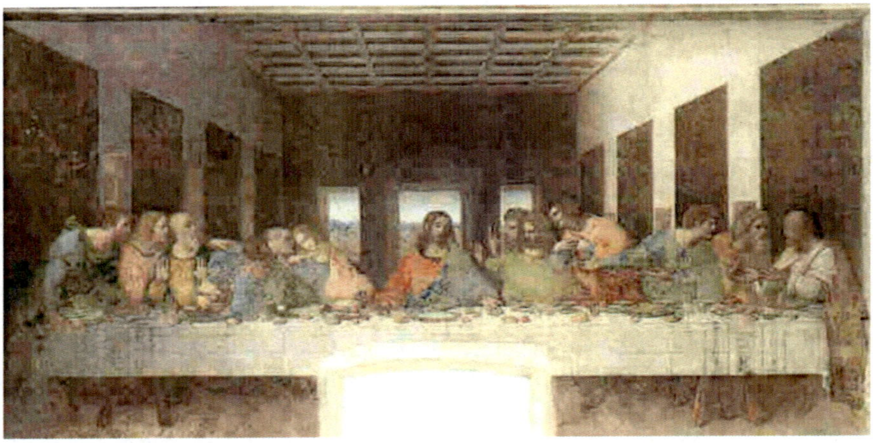

Fig. 4.1 Leonardo da Vinci the Lord's Supper in Milan, Santa Maria Delle Grazie (1495–1498) (Wikipedia commons)

the painting, he is listening to their answers.[4] His head is slightly displaced from the center of the door in the background, which relieves his face against the landscape. The upper border of the landscape defines a line on which the eyes of Christ are placed. Thus, the position of the central person in space is precisely calculated to match the geometrical and semantic centrality of the figure.

The geometrical arrangement of the thirteen actors in the scene has a fundamental symmetry—Christ versus twelve apostles, six sitting to his left, six to his right. The linear arrangement, which includes Judas, the traitor, is new. Most paintings, which treated the same topic before, have put Judas on the other side of the table such that he turns his back to the viewer (Fig. 4.1).

The morphogenetic organization follows basic laws of pictorial space, such as the laws of perspective discovered in the early Renaissance. In addition, it includes the model of regular polygons and polyhedra and the rules of symmetry. For example, in one of the preliminary sketches, John is resting his head on the table; the grouping of the remaining apostles on Christ's left is two plus three. On his right the grouping is one (Peter) plus two plus two. Leonardo restructures this scene to obtain geometrical symmetry, which may be expressed arithmetically: $13 = 3 + 3 + 1$ (Christ) $+ 3 + 3$.

In comparison with this order, we may observe that Giotto (1266–1336) painted his "Lord's Supper" with a significant non-homogenous group of three: Christ, John, and Judas on the opposite side of the table plus two groups of five apostles. The arithmetic order is $13 = 5 + 3 + 5$. Ghirlandajo (1449–1494) also adopts the central group of three persons, but he subdivides the five persons to the left. Leonardo's proportions are new concerning the tradition, and they define a new cognitive or image-schematic

[4] Luca Pacioli, Leonardo's friend, interpreted that Christ has just said: "Unus vostrum me traditurus est" and that Christ, at this moment, accepts the beginning of his martyrdom.

model, which introduces a new reading of the biblical story. Indirectly, his narrative interprets society and the relation in a human group (a family, a group of friends, a group of professionals, a political party with its leader, etc.). Therefore, Leonardo's organization of the topic creates a new frame of interpretation applicable to spiritual life and political and everyday affairs. In this sense, it can serve as background knowledge for innumerable communication situations. It is responsible for a piece of art's human and social relevance beyond its aesthetic function. Similar effects in modern media are analyzed in the case of "social semiotics"; cf. Kress and van Leeuven [15].

The utterance of Christ: "One of you will betray me", is a force, the effect of which makes a visible and emotional impact on the apostles. Like a shock wave, it hits most strongly the two groups sitting directly to the right and left of Christ and, to a lesser degree, the exterior groups. If we consider the nearer groups, James is pushed back, whereas John stays calm, although displaced relatively to Christ, and Judas is insensitive to the force emanating from Christ. He uses this moment to grasp the bread. These two groups are more agitated than the calmer outer groups. Thus, the dynamic effect of the words of Christ is represented as a wave with repercussions and vortices. If we analyze the single groups of apostles, their postures, and gestures, we can further decompose all four groups of apostles into two plus one (center). The central person neutralizes and stops the movement issued from Jesus and thus brings it to rest, the natural locus of motion in Aristotle's physics. It is as if the blow of the utterance had dynamically shaped the four groups and their subgroups. As a result of the impact, the grouping seems to be the natural consequence of the underlying dynamics. Judas is separated from the other apostles in his dynamics. He appears to be lost for any positive effect, holds his money in his right hand, and stops grasping for a moment at the bread. He shows a closed, sinister face which is in complete contrast to the face of John illuminated by the light from the left.

Leonardo tries to organize his composition to show the force's origin, its immediate effects, and the multiple structures created by the percussions of the power which came from Jesus. As the central force's emotional and intellectual effects are the painting's main topic, Leonardo reorganizes the scene's geometry to arrive at an optimal representation of the forces and their effects on body postures, gestures, and facial expressions.

The semiotic revolution in Leonardo's painting prefigures the scientific revolution announced in Copernicus' *De revolutionibus orbium* (1543) and brought to the summit in the work of Galileo and Kepler. Together with other intellectual and religious changes, they remodeled the common knowledge, the worldview of educated people in the sixteenth century, and Western civilization in the following centuries. The content of the painting emerges via an analysis of the natural physical and fluid dynamics. Leonardo can organize the semiogenesis of his mural painting in close relationship to the natural dynamics. It becomes a kind of "map" of the main contents of Christianism. It goes beyond religious standards in the second millennium and can code a universal human message.

The second painting, "St. Anne with the virgin and the child", belongs to the late period; Leonardo brought it to France towards the end of his life; it is now one of

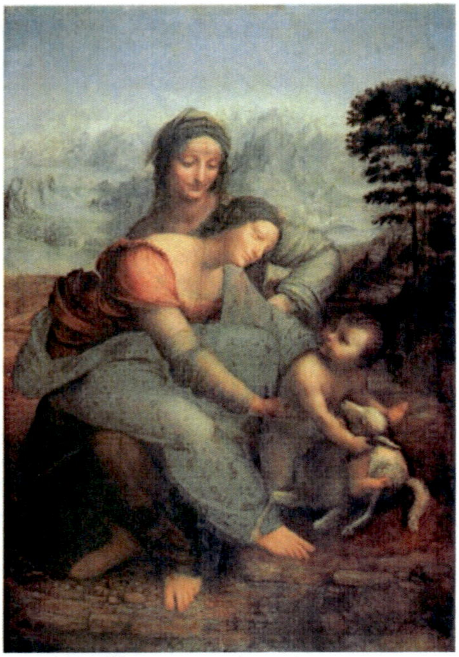

Fig. 4.2 Leonardo da Vinci: the Virgin and Child, with St. Anne (c. 1501–1519, Louvre, Paris) (Wikimedia)

the treasure pieces of the Louvre in Paris. In its center is the mother of Jesus, Mary, and her mother, Anne. Both care for the infant Jesus on their knees. In the chapter on myth and religion (Chap. 5 of this book), Mary is the second major attractor of Christianism. Thus, the two paintings analyzed here are like an introduction to the mythical message of Christianism (Fig. 4.2).

In the painting by Leonardo, three persons: St. Anne, Mary, and Jesus (of different ages and sexes), are readily identifiable as essential topics. Moreover, from the Bible, we know that there is a kinship relation: Anne—*mother o*f—Mary—*mother of*—Jesus (by transitivity, we know that Anne is the grandmother of Jesus, by inversion that Jesus is the son of Mary and Mary the daughter of Anne). This knowledge level already constitutes a second space related to the purely visual space that we construe as we look at the painting:

Anne—background to—Mary—background to—Jesus
Anne's head—*above*—Mary's head—*above*—Jesus' head
Anne—looks older than—Mary—looks more aged than—Jesus
Anne—*supports*—Mary—*helps*—Jesus.

Anne and Mary form one group (by proximity), and Jesus and the lamb another.[5]

[5] The colors help to complete the visual map. For example, in Leonardo's painting, both mother and daughter have a green mantle, and Mary can be seen to wear a red robe; both are subsumed under

The painting contains a rich geometric and dynamic structure (weights, bar centers, force lines, and gaze directions) used in many of Leonardo's works. A static representation would be insufficient for the painting's pictorial and narrative aims. Furthermore, this piece is typical of the concept of *dynamic valence*.[6]

In the case of this painting, we have on the surface a quaternary constellation: Anne–Mary–Jesus–lamb. Considering the force fields and actions, one notices that a primary interaction links: Mary–Jesus–the lamb.

- Mary *pulls on* Jesus
- Jesus *pulls on* the lamb
- The lamb *resists*
- Jesus *resists* being pulled away from the lamb.

There is a conflict between Mary, who tries to prevent Jesus from seizing the lamb, and Jesus, who notices this (he looks back at her) but resists her action. This triad constitutes a force field, which dominates the message of the painting. A first schematic representation introduces two vector fields with attractors:

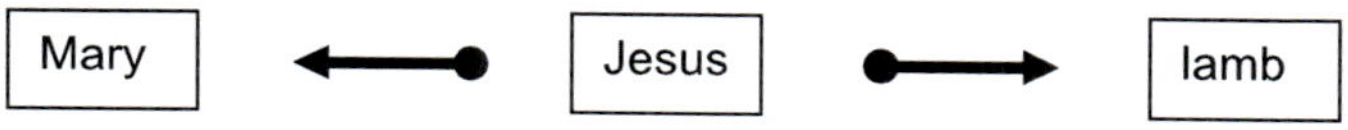

Jesus is in the metastable position between two attractors; the narrative (biblical) content of these attractors is.

- Mary: His mother. She cares for her child.
- Jesus: He feels the duty to sacrifice and to leave Mary behind.

The cognitive dynamics lying at the heart of the two paintings by Leonardo may be described in the context of dynamical semiotics (cf. Wildgen [34]: 68–72). The constellation of forces between Mary–Christ–the lamb corresponds to the fundamental dynamic schema of transfer (cf. Sects. 1.4 and 6.6 in this book or Wildgen 1982).

As the schema of transfer does not describe all the interactions in the pictorial composition, one has to consider two elaborations:

- Anne supports/anchors the whole event (physically and genealogically). She is a fourth attractor, sustaining the event on her knees.
- The manner of "transfer" is further elaborated in the painting and could be described in a sentence such as *Mary* ***trying*** *to prevent Jesus from seizing the lamb*. This sentence goes augmented by a modal verb (try to) beyond the elementary schema derived from catastrophe theory (cf. Brandt [2] for an analysis of modality in terms of catastrophe theory).

---

the color opposition: red-green. The lamb is a symbol of the sacrifice of Jesus and an attribute of John (Baptist).

[6] Peirce was the first to propose a general scheme on this level of abstraction. His monads, dyads and triads are considered as dynamical wholes not reducible by simple composition.

In the painting, one sees Mary's hands seizing Jesus and the hands (and feet) of Jesus holding the lamb, and we see that Jesus has a firmer grip on the lamb than Mary. This is because turning his head opposes the force direction of Mary's hands.

One could further distinguish two levels:

- An interpretation based on everyday experience with human interaction.
- An Interpretation linked to the Bible.

The distinction between the two levels of analysis is natural because in real-life situations, i.e., when one observes a mother interacting with her child, the first interpretation suggests itself. In cultural traditions that have elaborated a collective visual or linguistic memory (e.g., of primary biblical contents), the second interpretation is superimposed on the routines of the first one. It is typical of art to presuppose cultural knowledge and special "reading"-skills; i.e., the meanings go beyond everyday experience and are rooted in knowledge, the second pillar of semiotics (cf. Chap. 7).

4.3.2 The Morphogenesis of Abstract Painting in William J. M. Turner

Natural scenes and even actors may fade from sight, e.g., a landscape in fog or the transition to darkness. A person under similar conditions disappears also; in other circumstances, the viewer is only allowed a short glimpse, as when passing by on a train. Reduced visual cues are not necessarily *abstract* but may be *accurate*. Another example of *abstract* perception occurring in *real* vision may be found in cases in which people half-perceive or only implicitly perceive objects which appear in a repetitive or monotonous context. Examples are the telephone poles at the side of the road on a long car trip, the furniture of a room one passes through daily, and so on. Although viewers have the physical ability to perceive the whole situation in a precise manner, they largely ignore the multitude of details and pick up one or two specific features. In the same way, the artist may prefer to neglect all foreseeable and general characteristics. This process is analyzed in some paintings by William Turner.

William Turner (1775–1851) stands in the tradition of British landscape painting. Around the first half of the eighteenth century, aristocratic families practiced sending their young men on a "Grand Tour" to visit Europe and Italy. Paintings of scenes encountered on tour, mainly in the dominant Italian style, were brought home as souvenirs. However, painters like Wilson, Gainsborough, Wright of Derby, Cotman, Girtin, and others, departed from this tradition, and Turner radicalized the move against topographical preciseness (cf. Meyer [21].

The loss of referential distinctiveness and the neglect of identification of specific objects, cities, and landscapes may have been motivated by Turner's preference for vague surfaces under specific (natural) conditions. He felt uncomfortable working

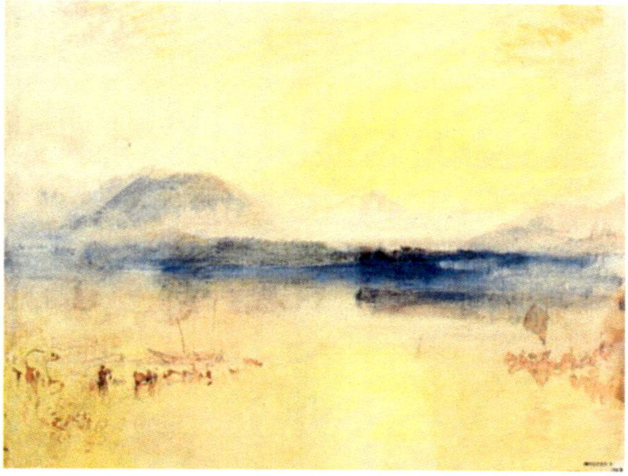

Fig. 4.3 Geneva: the mole, the lake, and the Savoy Hills, 1841 (Wikimedia)

with his materials out of doors and preferred a more economical or even minimalist method. For example, in a painting of the Castle of Chillon at the lake of Geneva from 1809 (British Museum), a precise representation with persons in the foreground, buildings, the lakeshore, and mountains is given. However, in a watercolor painting from 1841, the same lake of Geneva is depicted in a very vague fashion (Fig. 4.3).

In the second painting (see above), one can still recognize mountains, the shore of the lake, the water surface (blue), ships, and possibly people, but the symphony of colors, the transitions between surfaces, and indirectly the emotional values become dominant. Some later paintings reflect very specific, rare, and traditionally not represented phenomena such as "Snowstorm–Steamboat off a harbor's mouth making signals in shallow water, and going by the lead".

Turner gave this accurate description to avoid an interpretation of the picture as fantasy or caprice. In a certain sense, some of these paintings are even more realistic than those by Leonardo because they refer to concrete, personal perceptions of the painter. Their objects refer to geographically and historically distinct entities which can be identified as parts of the context of the painter's life, episodes of his journeys, and situations observed and remembered. This subjectivity refers to the life and the body of the author. It is probably the new message which made Turner a precursor of the impressionistic style in the eyes of later generations.

The next section will follow this line, which became dominant in the artistic directions of Expressionism and Cubism. The work of Picasso could be a good example, but the strategy followed is even more apparent in the sculptures (and sketches for sculptures) of Henry Moore.[7] In 1932, Henry Moore met Pablo Picasso,[8] Jean Arp, and other modern artists in Paris; in 1936, he joined o group of surrealists

[7] An example of morphogenetic structures in modern art in given in Chap. 2 of Caliandro [5]: "Paul Klee's Grey Point".

[8] An aspect of Picasso's work, his series of repeating and deforming well-known pieces of art, is discussed in Sect. 4.4.

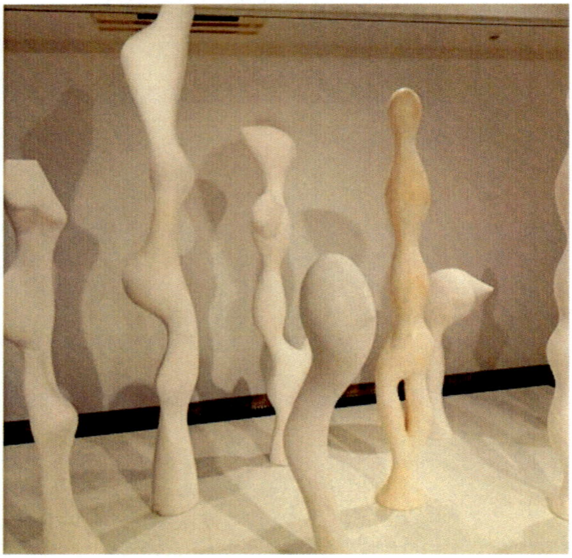

Fig. 4.4 Hans Arp (1886–1966) a selection of sculptures by Arp from the Sophie Taeuber-Arp foundation, Gerhard Marcks Haus Bremen, November 2022. Foto W.W

in London and began to produce abstract sculptures. However, the sculptures of the Swiss poet, painter, and sculptor Jean Arp (1886–1966) were already oriented toward biotic, natural forms, as shown in Fig. 4.4

4.3.3 Abstractive Reduction of Human Body Postures in Henry Moore's Sculptures

To restrict the scope of the analysis, we only consider the "Reclining Figure" topic, which is frequent in Moore's oeuvre. In a crayon drawing ("Reclining Nude", red and black crayon, ca. 1923; cf. Moore [24]: 96), Moore shows a realistic picture of a "Reclining Nude", a drawing from 1933 shows a series of projected sculptures on the same topic. Even if single parts of the sculptures may be recognized as belonging to a human body in a reclining position, these figures are like an exercise for variable shapes of persons in the given pose. What is left is the horizontality, the partial suspension (and support) of the body characteristic of a reclining posture (cf. Moore [24]: 112).

Henry Moore became famous for his large sculptures, many resembling human bodies, but some lack any referential support. In Fig. 4.5, one of many variants of Henry Moore's treatment of the topic in sculptures is presented and used for further analysis.

The head is a mass in the normal position (above the shoulders), and the direction the head is facing is distinguishable. Everything else: hair, ears, eyes, mouth is absent. The sharp frontal line on the "head" could be a generalization of the nose. The

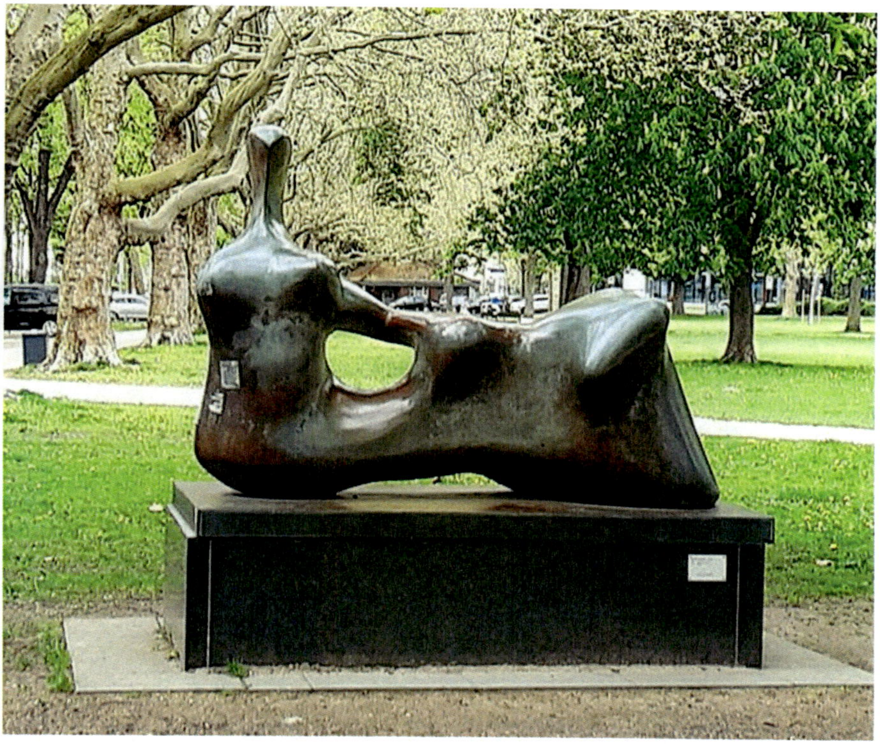

Fig. 4.5 Henry Moore. Reclining figure: Hand (1979) Moorweide in Hamburg-Rotherbaum Creative Commons CC0 1.0 Universal Public Domain Dedication

oversized extremities (feet) may be associated with the shape of covered knees in a reclining figure, and the large mass at the end could be a long cloth covering the open legs (cf. the drawings from 1948 in Moore's "Shelter Sketchbook", cf. Moore [24]: 151). The result of this "abstraction" can be explained by a series of transformations, which are visually reconstructed by the viewers as soon as they know the topic of the sculpture.

What is the prototype of a draped lying woman? The most straightforward scheme is one of two masses:

Mass 1: (head, trunk, arms)–mass 2: (legs)—transitory zone: back.

The transition zone is the most challenging part of the composition; it can be a sphere (maximally curved)—a bent cylinder (moderately curved)—or a regular solid, e.g., a pyramid, a cube, etc. All these solutions are at least implied in Moore's sculptures. Figure 4.6 gives an idea of possible compositions.

The narrative content of a "Reclining Figure" is concentrated in the verb: *to lie*, with its linguistic neighbors: *lie down, sit down,* and *stand up, rise.* The viewer of the sculpture may mentally add in masses for the head, shoulders, arms, elbows, knees, legs, feet, hair, eyes, nose, and mouth. He thus goes back on a path of abstraction

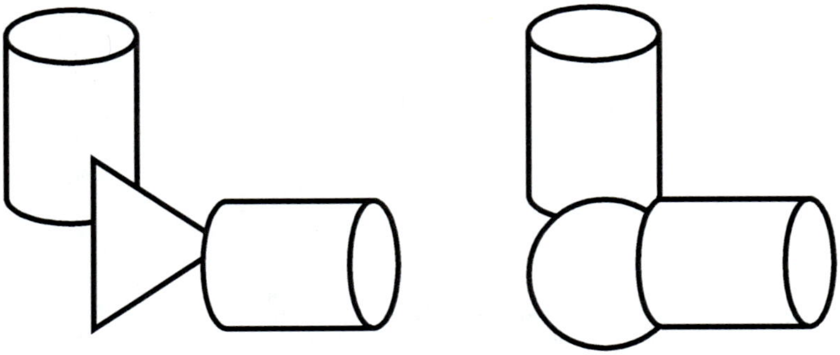

Fig. 4.6 Reducing the "Reclining Figure" to a cylinder, sphere, and (curved) pyramid

analog to that taken by the sculptor himself (or even engages in a longer "story" of artistic development). It may be understood as the trajectory from a prototype (or a set of prototypes) to our visual representation of a particular human body. Thus, the sculpture represents this cognitive itinerary (cf. Leyton [17]).

A more radical visual analysis of the human body destroys its unitary gestalt, its topological coherence. Two trajectories may be observed in Moore's oeuvre:

- Instead of one body (from head to feet and hands), several sculptural entities which together "represent" the "Reclining Figure" are used. Moore distinguishes two-piece, three-piece, and four-piece reclining figures.[9] The geometrical analysis is shown in Fig. 4.6 is insufficient to cover all the variants of the topic. Still, if one adds curved (and conical) cylinders, spheres with excavations, and bent cylinders, then most of the forms may be classified on this basis. The bending of lines, the curves, and the concave or convex surfaces refer more specifically to the realm of biological entities.
- Beyond convex and concave limbs and surfaces, one can observe gaps. They may appear in a reclining figure, e.g., between arm and body, or arms and legs, and the support.[10]
- In other sculptures, the title of the sculpture abandons the reference to human bodies, as in: "Three rings, 1966–67" (Moore [23]: Nr. 74).

Instead of continuous (differentiable) curves, sharp edges ("knife-edge"), pikes, and strings attached to bodies appear. The combination of smooth surfaces, knife edges, and pikes is illustrated in the sculpture: "Three points 1939–0" cf. Moore ([23]: 45, Nr. 37).

---

[9] In Moore [22], examples are shown in Nr. 10 (one-piece), Nr. 11 (two-piece), and Nr. 40/41 (three-piece); the four-piece composition (Nr. 86) is not called "reclining figure" but visibly belongs to the same family of sculptures.

[10] Cf. Moore [23]: Nr. 68, 71 "Reclining Fig. 1939", Nr. 73 "Reclining figure (external form) 1953–54".

4.3.4 Conclusion

Innovations in art history are more than temporary changes in taste or dominance shifts in schools' rivalry. They contain at their heart a "problem" and its "solution" by the artist. This feature has been shown in the case of Leonardo. The arrangement and the postures in the "Lord's Supper" contribute to the visualization of the dynamics inherent in the narrative episode represented in the fresco. In the series of works dedicated to the triad of St. Anne, Mary, and Jesus, a dynamic configuration with attractors, repellors, forces, weights, and counterpoises is visually organized together with a system of view lines. The case studies of Turner and Moore showed the natural process of symbolic abstraction in landscape painting (Turner) and figural sculpture (Moore) and the directions of their gradual development in the artists' oeuvre. The activity of the artists is a kind of artificial morphogenesis. It applies natural dynamic schemata known from nature and thus gives genuine meaning to the artifacts created. The viewers can recover these meanings due to their own experience with nature and the fact that they are also part of this nature.

4.4 Secondary Morphogenesis or "Meta-representation" in Art

The label "meta-representation in art" has at least two content levels of representation (cf. Wildgen [38] for more details). First, it can be shown in the case of a classical painting by Diego da Silva y Velasquez (1599–1660), "Las Meninas". This famous painting by Velasquez is discussed extensively by Foucault in the first pages of his book: "Les mots et les choses" Foucault [9]. It is shown together with one (of several) meta-referential paintings by Pablo Picasso in Fig. 4.7.

In the original picture by Velasquez, we can find at least three parties looking at the scene, which is the central concern of the painting, i.e., the princess escorted by her ladies on the left and right and accompanied by two dwarfs and a dog. The person standing behind the door and looking at the scene looks simultaneously at the king and the queen. They appear in the mirror and are thus only represented indirectly (and with side inversion). The painter, Velasquez himself, also looks at the monarch pausing his work for a moment. One may infer that, when he continues his work in one moment, he will again look at the scene with the Meninas and his painting. As he is probably not painting the back of the princess, she and her environment must turn around to continue this painting session. It is evident that the painting on which he is working is not the one we see, but one where either the princess is represented or a different one. In this case, the visit of the princess would be accidental.

Picasso's painting reproduces Velasquez's painting under deformations (58 paintings on canvas were created between August and May 1957 in Cannes). Instead of starting from a real-world situation like Velasquez (even if it is challenging to infer the details of this situation), Picasso takes the famous painting as his starting point,

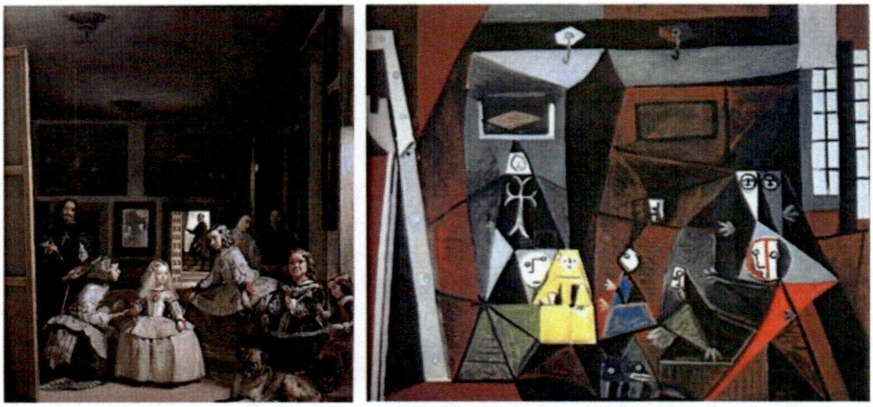

Fig. 4.7 Las Meninas by Velasquez (Museo Nacional del Prado) and one (of 58) meta-referential paintings by Pablo Picasso (Museo Picasso, Barcelona) (Wikipedia and Wikiart common)

creating a whole series of cubistic reinterpretations of this painting. His art is meta-representative because the viewer can guess the original topic and focus on Picasso's metamorphosis of the subject and his visual comments on it. Moreover, Picasso's characteristic style is a signature of his presence in the artwork.

In Velasquez's painting, the mirror plays a significant role. This meta-effect can be enforced not only in the painter's self-portraits but can also be recursively applied. A good example is the painting by Salvador Dali's (uncompleted) stereoscopic painting "Dali de dos poignant Gala de dos" which shows Dali seen from behind looking at his wife, Gala, who sits in front of a mirror. Both are shown from behind and in front (in the mirror) (Figueras, Fundacion Gala-Salvador Dali; cf. Descharnes and Néret [8]: 201). The action of the mirror could be repeated whereby the painter and his model would appear not only twice but three, four, and many times. As in the case of video feedback, repeated mirroring can have a chaotic attractor and destroy the original input, resulting in fractal patterns analyzed by Benoît Mandelbrot (1924–2010).

4.5 Fractal Art in Pollock and Art as Social Sculpture and Diagram in Beuys

The last section has shown that meta-representation in the art can drive morphogenesis to a chaotic result. In Jackson Pollock's art (late period), the fractal character of his painting is the primary goal, the actual content. All narrative elements (referring to texts as in Leonardo or indoor situations as in Velasquez) and all figurations based on spatial and temporal experiences are eliminated. What is left is the configuration of points, lines, surfaces, and colors.

4.5.1 *Chaotic Semiogenesis in the Art of Jackson Pollock*

"No chaos, damn it" replied Jackson Pollock (1912–1956) when critics described his art as chaotic and unstructured. However, even before 1947, when Pollock began dripping liquid paint from buckets onto the colossal canvas lying on the floor, some well-known artists sprayed paint on surfaces or even shot with paint guns at them. Strict, even linear patterns had initially dominated non-representational art; an explosive variety of forms from Dada to Surrealism replaced them. But what does "chaos" mean in art? Is it an accusation that artists have to defend themselves against, or can chaos in art be the new kind of order?

Originally, chaos meant a situation preceding creation, the chaos before the first day of creation. Similarly, the end of the world or the deluge becomes the epitome of chaos, and many classical painters have illustrated this situation. For example, Leonardo da Vinci depicts rushing floods (cf. Wildgen [41]: Sect. 3.2); in other depictions, people panic in all directions. But chaos is also, and in recent decades primarily, a mathematical concept. The core meaning of chaos has three facets:

1. The simple order of point, line, surface, and body (dimensions 0, 1, 2, 3) can be broken up if fractional dimensions (fractals) form the starting point. A line is interrupted infinitely often (Cantor line), a surface is "cramned full" with lines and curves, and a body is perforated infinitely often (Menger sponge). The dimension of these formations can be calculated between 0–1, 1–2, or 2–3.
2. Instead of being straight or smooth, a curve or curved surface can be so "crumpled" that it does not become smooth even at infinite magnification. In this way, Mandelbrot simulated mountain contours and coastlines or the shape of clouds using fractal models.
3. Events can also be chaotic if the number of alternative states increases drastically; the Feigenbaum diagrams are characteristic of this. After the Big Bang, the universe branched out almost infinitely. Structures with few branches, like trees in nature, or the structure of sentences (constituent structures) are special cases.

The fractal dimension of Pollock's paintings can be calculated using the box-counting method. In this procedure, line or empty space occurrences are counted in ever smaller sections of the picture. It was found that between 1943 and 1952, i.e., the period in which Pollock's dripping technique developed, the fractal dimension increased from 1 (1943) to 1.72 (1952). Analysis of the films showing Pollock's painting indicates two characteristic sizes. First, in the range from 1 mm to 5 cm, the fractal pattern is caused by hand movements while dripping. Second, in the range between 5 cm to 2.5 m, they follow the arm and body movements around and into the canvas. Mathematically, multiple pendulums (e.g., combined limb movements or limb and brush movements) have chaotic phases, which explains the fractal patterns. However, this characteristic only becomes clear because Pollock blocks the usual left–right reading of the picture and avoids centering or hierarchies. As a result, the visual intensity is refocused; the experience is relieved of recognizable ritual patterns and opened to new impressions.

Pollock's first large-scale picture focused on lines, was a commissioned work for Peggy Guggenheim from 1943: Mural (247 × 605 cm). The dripping technique, the reduction of color, and the all-over pattern distribution are characteristics of the picture "Number 32" from Pollock's main creative phase (1950 (North Rhine-Westphalia Art Collection, Düsseldorf, Pollock-Krasner-Foundation; visit https://www.kunstsammlung.de/de/collection/artists/jackson-pollock). The following properties stand out:

- Monochrome black paint, no title. Before the dripping phase, Pollock had taken great care in choosing a title.
- There are no contoured areas in the picture, only the painting ground, hairlines, drips, thickened spots, and emphasized curve segments.
- There are no references to objects (human, animate, artifacts) or forms that can be interpreted as such, and no conventional symbolic meanings.

But what "meaning" do these graphic structures have? First, the missing properties make it clear that no external world (with top/bottom, depth effect) and no surfaces with light and shadow are represented; the picture shows no reference to what has been seen or recognized. Secondly, it refers to the process of production, the artists' movement patterns, and thus to the person in the act of painting. The image "tells" itself; it is self-referential, a central feature of chaotic patterns. Thirdly, it concentrates on the texture, the graphic, and the calligraphic surface structure.

The pictures are subject to an organization in which the artist's aesthetic decisions determine the direction but cannot be strictly separated from the self-organization in the painting process. This feature points to the "écriture automatique" introduced by André Breton. In art, the phases of chaos can have a liberating character and make undreamt-of possibilities available. After that, however, the products of these phases must be integrated visually and cognitively/emotionally in a field accessible to a larger public of art lovers, artists, and art collectors.

4.5.2 Social Plastic, Diagrams, and Morphogenetic Fields in Beuys' Artwork

Joseph Beuys had only begun his career in the late fifties of the twentieth century when Germany reappeared as a global player in the international political and cultural scene. Contrary to Jackson Pollock, Beuys was himself a theoretician of art engaged in cultural politics and a protagonist of the rapidly changing avant-garde art scene in Europe and America in the sixties and seventies (cf. for a detailed analysis of the art of Beuys [42]).

In the sixties of the twentieth century (first in 1967) and the seventies, Beuys began to enlarge and redefine his concept of sculpture and art and created the idea of "social sculpture" ('Soziale Plastik'). The materials are not only of organic and animal origin but also social and political (society, humanity). Thus, Beuys expanded the type of materials for "sculpture":

- From stone and steel to "warm" materials (wax, honey, grease, animals)
- From animals and humans to social and political entities and even to humanity.

Consequently, he sketched the program of a new political movement that, together with other currents, contributed to the foundation of the "Green party" in Germany. Beuys even stood on the list for the first elections in which they participated. He also initiated a program for alternative universities. Thus, the "Free International University of Creativity and Interdisciplinary Research" was realized in Ireland (Dublin, Belfast, and Derry).

The innovative power of art meant a new perspective beyond politically and intellectually proposed alternatives. Symbolic messages should avoid the deceptive strategies of art depending on political trends (neither affirmative nor in opposition to them); they should instead be non-directive, bodily rooted, and in continuity with nature. In a broad sense, the art of Beuys is morphogenetic; it reduces the elite gesture of artists, which is academy-affirmative and a-social, a-political.

For Beuys, speaking corresponds to a kind of shaping of ideas and thus is akin to sculpture. Language as art is, for Beuys, primarily the play with language sounds and underlying concepts. Its goal is the creation of new social meaning. In this line of thought, art is a conceptual procedure (an operation with concrete forms, visual or spoken) and gives access to unknown pictorial and abstract spaces.

Beuys combined figurative drawings with graphical symbols (lines, circles, arrows) and written words in diagrams performed on blackboards or the floor. In a diagram of this period (1972), one finds several scales of concepts:

- Chaos–order; birth–death; cold–warmth; volition–feeling–thinking (Chaos–Ordnung; Geburt–Tod; Wärme–Kälte; Willen–Gefühl–Denken); Beuys indirectly points to the Aristotelian technique of contraries.
- Order and form are depicted by crystals (regular solids). This reduction to primary, ideal forms is a traditional topic since Plato's dialogue "Timaeus". In the Platonic tradition, Leonardo, Dürer, Kepler, Goethe, and many others referred to regular solids and crystals to explain regularity in nature. The opposite—chaos, turbulence—is represented by spirals and chaotic attractors. Different drawings/diagrams allude to mathematical forms in nature and art and concepts in the natural sciences.

On a more general level, the lesson taught by modern art says that the human experience of art is, on the one side, very open and can be adapted to new experiences. On the other side, it is guided by social processes, be they local such as by groups or centers of art, or global such as the cultural changes after the wars (1919 and 1946). In the work of Beuys, the dimension of reflective processes, art theories, and art politics (even of a utopia where artists shape society or humanity) became a prominent factor. Meanwhile, this trend may have lost its fascination, but it can be expected to show up again under specific conditions. Permanent change seems to be the distinctive mark of art in the modern world.

4.6 Morphogenesis in Human Settlement and Architecture

The prototype of a shelter in human prehistory was a cave not far from the shore, a lake, a river, and possibly on the slope of a hill for a good overview of the surrounding area. This site will be found along the migration routes of Homo heidelbergensis and Homo sapiens (following shores and rivers) and in the caves which document the social life of Homo sapiens. The prototype for the (primarily seasonal) dwellings had the following zones:

- Cave entrance (often under some rocky roof) with an external zone for social life, tools manufacturing, and distribution of prey and harvest products.
- Cave mouth (still with light from the entrance or fire) for family life and sleep.
- Deep cave interior for spiritual events like rituals.

The primary opposition: abris or shelter versus area of chase/harvest, including shores, lakesides, or river banks, defined the proximate ecology. The shoreline could be extended via migration, i.e., the next generation could move to a similar ecology along the water's edge. This migration line defined the first transport technology: the raft, which expanded the natural capacity of wading, swimming, and diving. The boat became the archetype of instrumental locomotion. Until the Neolithic revolution, all settlements were seasonal because humans had, like other carnivores, to seasonally follow their prey or, if they were scavengers, the predators. This pattern continued until recent times in many parts of the world.

Instead, more permanent settlements appeared when agriculture became dominant.[11] With the permanent settlement, building a house became a specific goal and led to the advent of architecture. Some aspects of morphogenesis in architecture are the topic of the next section. Architecture's long and diverse history is neglected, focusing on specific phenomena where the morphogenetic aspect is manifested.

4.6.1 Morphogenesis of Urban Structures

The first towns emerged, e.g., Jericho at 8,000 BC, Çatalhöyük (7,500 BC), Eridu, Ur, and Byblos (between 5000 and 4,000 BC). These 'towns' already contained a large population and had a wall that limited the town area and later provided defense. Particular derivations of the abri (cave) and the town are burial sites, graves, and graveyards (grave villages and towns).

The morphogenesis of the town is governed by two forces that define its shape:

(1) The individual houses and their use by the owners/builders.
(2) Collective protection (against enemies) and communal functions (market, political decision, religious practice).

[11] In the case of an agriculture based on burning brushland, the settlement moved slowly in the controlled area.

Fortification was necessary for the accumulation of goods and wealth in Neolithic societies. However, this development also led to social segregation inside the town and corresponding differences in house building and defense style.

The next stage is reached when the local type of self-organization, which is only globally controlled by the protective wall, is replaced by a central plan. This can be shown in the case of rectangular grids in towns. The first grid pattern in towns probably emerged in the Indus Valley, and Herodotus reports that Babylon had ample straight avenues intersecting at right angles.

4.6.2 Self-Organization of Urban Structures Versus Ratiogenetic Planning and Imitation

The Romans, who exported a standardized town plan to many regions of their empire, developed the pattern they had found in the Greek colonies in Italy (cf. Naples). The Roman Paris also showed such a grid. In the Middle Ages, this traditional pattern was lost in Paris but recovered in the thirteenth century in France ('ville-neuves' and 'bastides') and England (Welsh 'bastides'). After its destruction by an earthquake in 1755, Lisbon was reconstructed on a rectangular grid pattern.

If the early or medieval towns show a compromise between more or less free individual settlements and a defensive fortification (ideally in a circle), the new towns planned by a centralized agency demonstrated the will of one person or an administration. On the other hand, the idealized geometrical forms had to respect a specific population's geographical, climatic, and economic conditions. Therefore, the outcome is mostly a compromise first between the *individual* goals and values and *collective* ones (often represented by some institution or leader); second between specific *natural* conditions that change in time and *traditional* forms that have gained some credit (fame).

Upon this rather basic substratum, several symbolic levels could be grafted. They communicate differences in power and wealth or religious/ideological values. The question of how such compromises are reached is difficult because many alternatives may exist, and chance factors may influence the outcome. This situation can be modeled dynamically in *synergetics*, which analyzes the combined effect of many subsystems and focuses on simple rules which can reduce their complexity. Another choice is *chaos theory* and *fractal geometry*, where irregular patterns can be understood more easily.

The main differences between architectural and urban semiotics relate first to space and time. Spatially, the city not only includes residents and guests, but it also includes houses, streets/squares, boundaries (fortifications), and possibly a river or an estuary. Just as the house is embedded in a landscape structured by paths, fields/gardens, forest/lake/moor, the city is also embedded in a network of trails, often in a web of trunk roads. The house contains a family (or more) and their property. The city includes a population. The house has a prominent space as its center (e.g., the living room), and the city has a building or a central square surrounded by buildings as its center. In terms of time, a house can be designed and built by the owner and his architect. However, the morphogenesis of a city can last for centuries, even thousands of years. Existing spaces are constantly being rebuilt, often while preserving their essential functions and meanings. In this temporal perspective, the city is relatively independent of individual planning. In cases of destruction through conflagration, war, or brutal urban innovation breaks, this quasi-biotic morphogenetic city becomes a giant piece of architecture or even a monumental sculpture (see the capital of Bresil, Brasilia, planned in 1956 and founded in 1960).

Beyond architectural and urban morphogenesis lies the morphogenesis of landscape. The landscape, e.g., of a country estate in Veneto (see the analysis of the Villa Barbaro in Maser by Palladio in Wildgen [41]: Sect. 8.3), is reflected functionally and symbolically in the architecture of the country house (estate). Seen globally, many courtyards, outbuildings, fields, paths, forests, and waters result in an encompassing whole, e.g., the villa formed by an architect like Palladio. The landscape's geological formation, particularly evident in the case of seashores and mountains, belongs to another type of morphogenesis largely independent of humans.[12] For the artists of the nineteenth century, this landscape was, therefore, fascinating in its natural symbolism. The romantic painter was able to illustrate elementary forces and contents in the painting of a landscape. However, to the extent that tourists took hold of exciting landscapes, e.g., the high mountains, the natural state of these landscapes was increasingly destroyed and changed by countless (often architectural) interventions. Ratiogenetic forms conflict with natural morphogenetic symbolic forms, either those emerging independently in our ecology or those formed by the natural dynamics of human life.

4.6.3 *Political Ethics and the Architecture of the Town Hall in the Late Renaissance*

Inhabitants or tourists may value a room, a house, or a street/quarter. Their subjective choice can influence the price of apartments or hotels. Therefore, urban dynamics can be understood as a search for a value maximum. This search is limited by the costs and has to respect the hierarchy of values accepted at the construction time. Thus the kings or the archbishops can have the privilege of the largest, highest, best-decorated palace or church. On the other hand, an entrepreneur may be able to

[12] However, the natural structure of a landscape can be also artistically formed, for instance, by the landscape art of Christo (1935–2020), see for example the Veiled Coast (1969, Australia) and the Valley Curtain (1970, Colorado), Land art reacts to the post-industrial (transformed) landscape; see Paquet [27].

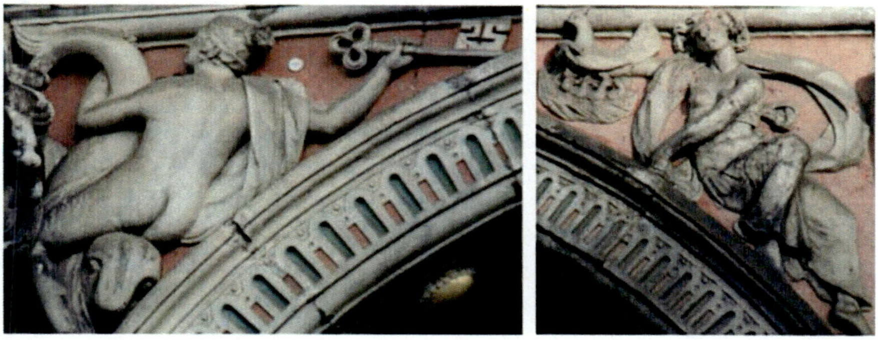

Fig. 4.8 Allegories of "temperantia" and "vigilantia" on the arcades of the Renaissance town hall in Bremen (Photo W.W.)

buy all the houses in a particular street and rebuild it, thus imposing his architectural imagination.[13] Dramatic value changes occurred during the Reform (sixteenth century) in Northern Europe, when monasteries were abolished, i.e., either destroyed or transformed for other institutions. Beyond this destruction, new architectural forms visualize the values of the reformed religion. Many are ambivalent, can support different values, be re- and misinterpreted, and may coexist for extended periods. Thus in Bremen, the massive Romanic/Gothic cathedral exhibits the self-evaluation of medieval archbishops. In contrast, the smaller but still representative town hall stands for the values of the commercial town with its elected town council. The value system of the commercial city, as a member of the Hanse (north European commercial league), is expressed by the oversized statue of Roland (with sword and gold-braided coat, built in the year 1404). It represents the town's independence from the feudal regiment (the archbishop who rules the whole territory surrounding the free Hanseatic town). The Gothic decoration of the town hall was refurnished around 1610 in the style of the Northern Renaissance. The figure of a woman riding a fish in the arcades of the town hall with a key in her hand is often interpreted as an allegory of Bremen, i.e., "Brema", because the key in her hand is the emblem of the town. However, this sculpture reinterprets contemporary prints and can be reconstructed as an allegory of "temperantia", i.e., temperance, moderation, a virtue political leaders should demonstrate. A similar misinterpretation occurred in the case of the allegory of "vigilantia" (vigilance), which became the center of a myth of origin invented in the nineteenth century. The first citizen may have chosen this place when they observed that a hen protected her chicken on this dune in security against inundation (cf. for more details Wildgen [41]: 274–284). These examples show the easy transformation of an "interpretant" given architectural sign forms with a forgotten historical context (Fig. 4.8). The invention of a myth of origin may even occur ad hoc in modern times (see Chap. 5 for the dynamics of myths).

[13] Thus the coffee-entrepreneur Roselius bought the old houses in central Böttcherstraße in Bremen and built a neo-classic street, one of the major tourist attractions of present Bremen.

Thus, if in 1400 the figure of a warrior, a henchman of Charles the Great, represented the values of an ambitious commercial town, independent from local feudal sovereigns and with "free" citizens, the copious illustrations on the late-Renaissance facade demonstrate the erudition of its council members and their principles of governance.

Beyond basic value systems linked to dominant parties in the township, architectural styles shape urban architecture. For example, after the destructions of World War II in Germany, a conflict between historical reconstructions aiming at continuity with traditional values on the one hand and new value systems linked to modernization on the other triggered intense public debate. Other conflicts involve local value systems (neighborhoods) against global plans for traffic flow. These examples show that the landscape of value-ascriptions is multifaceted, if not fractal, i.e., many small and large value systems coexist, and a complicated bargaining process takes place. As a result, urban developments may be amplified or stopped by these processes.

The specific features of value dynamics are of two kinds: Firstly, they exhibit a strong *memory effect*, e.g., those values valid for earlier times and generations are conserved with preference and value changes are suppressed, i.e., the dynamics are conservative. Under exceptional circumstances (large natural catastrophes and political revolutions), a rapid value change can overwhelm such barriers and trigger sudden changes. Secondly, there is an "echo effect". If a value-based structure is destroyed, the void tends to be filled by forms with similar values (of similar type). An example of the first tendency is the reconstruction of historical city centers in Western Germany after the bombing of World War II, which destroyed 70 to 80% of the buildings. As a compromise between memory and modernization, an eclectic style emerged. An example of the "echo effect" is the instauration of buildings with a cultural function in places where earlier religious architectures were destroyed (for instance, in the time of Reform). The museum or theater is a morphogenetic "echo" of religious buildings with spiritual, intellectual, and artistic values in earlier times.

4.6.4 Biomorphic Principles in the Architecture of Antoni Gaudi

The still unfinished project for the Sagrada Familia Cathedral in Barcelona is an excellent example of how morphogenetic aspects surface in urban architecture. Antoni Gaudi (1852–1926) proceeded from the principle that what is beautiful and authentic in architecture is not invented by man but is found by man through observation and imitation of nature. The organic world and the principles on which the organic is based play an essential role in his architecture. In technical realization, Gaudi relies on principles of geometry and statics. However, the focus is not on the geometry of the land surveyor but on the geometry of living nature. Gaudi, therefore, experimented with convex and concave surfaces, with conoids, hyperbolic and parabolic surfaces, and their intersections. Since the mathematical mastery of these new forms was still insufficient, Gaudi tested the statics and dynamics of such constructions using his inverted rope models hung with weights. Currently, the vaulting of the Sacrada Familia has been completed using modern computer-aided design techniques (Fig. 4.9).

The towers of Gaudi's cathedral look like termite mounds. The nave, which was only completed before the sanctification by Pope Benedict XVI in 2010, reminds the visitor of a massive forest with high trees, the ceiling interrupted by windows, again comparable to spots of light penetrating the roof of a forest.

Other constructions copied from living nature are the roof construction of the opera building in Sydney in analogy to the spherical shell geometry (design by Jorn Utzon 1957). Unfortunately, the building could only be converted into a technically feasible construction after lengthy experiments (end of construction in 1973). This trend continues in today's virulent "soft" design based on computer simulations and computer aided design (CAD), which brings the architecture close to a sculptural style. Classic examples are the Guggenheim Museum in Bilbao (Frank O. Gehry, 1997), his Experience Music Project, Seattle, 2000, or within the Millennium Dome (London) the Body Zone, a structure in the shape of a reclining woman (Branson Coates, 2000); see Jodidio ([14]: 60f. and 32f.).

The consistent penetration of the construction with a biomorphic concept still distinguishes Gaudi's buildings since he designed the outer facade, the interior, and the furniture according to the same principles. Materials and building techniques are essential in searching for new forms. Examples are The iron constructions of the late nineteenth century (prototype: Eiffel Tower), the concrete constructions (Le Corbusier, Ronchamps Church), and stretched tent constructions (Olympic Stadium Munich and tent-like construction in Astana; Kazakhstan). Spider webs and tents as primitive (prehistoric) building constructions are implicit references to biological and behavioral morphogenesis. In all these cases, human construction implements

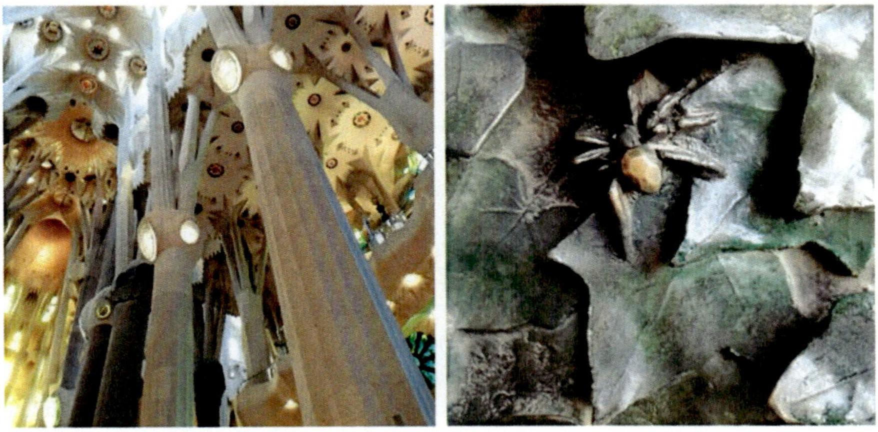

Fig. 4.9 The tree-like pillars and botanic/zoological details at the door of the cathedral "Sacrada Familia" in Barcelona (Foto by W.W. 2022)

constructions found in nature with different and technically improved means. The biological sense of natural construction has its roots in the instinctive behavior of an animal and, ultimately, in the evolutionary selection conditions of the species. Thus, at least in part, the meaning of architecture to humans is grounded in fundamental meanings in nature, i.e., it can be derived from biosemiotics (cf. Nöth [25]: 254).

The buildings mentioned above usually form nodal points in social life. For example, many people gather in the church, the opera, and the stadium. Here they perceive themselves as a community and take part in ritual acts. The simple dwelling houses, on the other hand, do not have this collective meaning; in return, they determine the private living environment and the intimacy of the residents and are, therefore, significant for families and individuals.

4.6.5 The Morphogenesis of Typical Harbor Towns

In the following section, urban structures focusing on harbor towns will be considered, i.e., towns on a river or estuary with access to the sea. For example, Paris on the Seine (more than 100 km from the sea) and Bremen (60 km from the sea) are cities on the river and traditional port cities. For the morphogenesis of harbor towns, cf. Wildgen [41]: Sect. 9.1 (in German) and [40] (presentation in English).

The primary feature of a harbor town is the opposition: land (shelter) versus water (motion). It implies using boats and ships, i.e., boats arrive and depart, ships are built (repaired), and goods are transported and commercially exploited (stocked, sold). A marketplace emerges, and eventually, factories and habitations for manufacturers are built. The stockpiling of goods required protection (against thieves or pirates). The emerging social stratification induced spatial differentiation (differences in housing) or even segregation (in city quarters). The value system gave rise to economic standards valid in a town/region and to exchange patterns between towns/regions (see the origin of banking). The rights and privileges of different groups or even of towns in a commercial network needed laws, courts, and administrations. The semiosis of the commercial system led to very complex rule systems, which eventually collapsed when the global conditions (of power or the available commercial routes or resources) changed. What remained stable over the centuries were the primary significance of land/sea and the global ways of exchange depending on the geographical and physical conditions of a region/country. In the following, we consider the aspect of architecture and urban structure as the outcome of semiosis in the (former) harbor towns: Lisbon, Paris, Bremen, and Aarhus.

Morphogenesis presupposes that there is a 'germ' of the town and that basic dynamic laws control its unfolding. In the case of harbor towns, we consider the phase transition of land to water and a singularity of paths. Prominent singularities are crossing two roads (or a road and a river) or the existence of an island. In the following examples, we have a shore or the bank of a river (with access to the sea) and a simple or double singularity of paths:

1. Lisbon: Two creeks flow to the shore between higher banks shaping a plain (actually the Rossio). The river opens to the sea.
2. Paris: An island in the river Seine leading to the sea and several hills around a flat and wet area. A road crosses the river via the island (Île de la Cité).
3. Bremen: A dune along the river, a tributary river forming an island, and two long-distance paths/roads, one crossing the river near the island.
4. Aarhus: A shoreline and a river mound. Neighboring heights delineate a flat area.

The geographical or geological situations may have had different types of affordance, i.e., the physical context allows or "invites" specific usages and functions in the perspective of the human actor (cf. for this term Gibson [10]). The affordance differs for alternating populations on the site and can change dramatically over the centuries. What counts in the case of harbor towns is the catastrophic transition: land to water, road connection (transport of goods and persons), and the shelter type (dune against floods, island against invasion, steep hills for defense). In the case of a harbor town, the transition line between water and land is an affordance for ship-berthing and shipbuilding. Later, the opposite bank and other places toward the sea can be used as harbor expansions. Finally, secondary functions like storage, manufacture, and market unfold in the neighborhood of the harbor zone. The evolution or unfolding of affordance over the centuries was different in the four towns compared:

(a) In Lisbon, many different cultures used the location (cf. Marcos [19, 20]), but the wet plain between the high grounds and the bank became, after its drainage, the center of the harbor (shipbuilding) and commerce. Steep fortified hills afforded its protection. In addition, religious functions were established on the heights.
(b) In Paris, the pre-Roman inhabitants of Lutetia probably used the island in the river as shelter and protection. The Romans built their town on the heights of the Southern shore (today Quartier Latin). The central road ('cardo') of the roman city corresponds to the boulevard St. Michel and the streets of the 'quartier Latin still visualize the roman grid (cf. http://paris-atlas-historique.fr/38.html). The river Bièvre and the zone around the bridge to the island in the south were the original harbor sites (cf. Velay [31]: 1–2).
(c) Bremen developed following its Christianization (it became the seat of a bishop in 787). The church on top of the dune may have been the decisive symbolic germ. With the medieval traffic on the North Sea and the Baltic, its small harbor evolved into an important commercial place in Northern Europe.
(d) In Aarhus, a Viking town was built on a triangular surface closed on two sides by shore and river. The third flat side could be artificially closed by a moat. Later Christian missionaries built a church outside this area which subsequently moved into the center.

The morphogenesis of a town is a kind of semiogenesis insofar as the physical factors: land/water–hill/plain, become influential forces only as they gain significance for the human settlers. Nevertheless, the governing forces remain so basic and stable that we can observe regular patterns analogous to morphogenetic patterns in physics, chemistry, and embryology.

Due to different ecologies and populations, European harbor towns differ in many respects. Therefore, the four cities compared can be classified into two groups according to their networks and areas:

- The north European network is linked to Viking trade routes (around 1,000 AD) and the Hanseatic trader league (1200–1500 AD). Aarhus and Bremen are examples of this evolution.
- Paris and Lisbon are mainly influenced by the urban patterns of the Roman Empire (its colonies), the medieval Arabic and Christian reorganization, and modernizations since the seventeenth century. Both towns were either the model imitated by modern European capitals (Paris) or were exported to overseas foundations (Lisbon).

The historical change of the four towns also influenced their architecture:

1. In Lisbon, the Visigoths fortified the steepest hill, which was further fortified after the Arabic conquest (719), to become the "Alcáçova", and after the reconquest (1147), the fort S. José. The Arabic mosque at the center of the Arabic Almedina became the (rebuilt) Christian cathedral Sé. Large monasteries and churches were placed on top of the neighboring hills, thus occupying the significant ("prégnant") hills punctuating the urban pattern. The commercial city moved to the flat part (former river bed) and the banks of the river Tejo (Praça de Commercio). From the shipyards on the Tejo, the expeditions to Africa (1419) and India (1488) opened the era of colonial expansion. An earthquake and tsunami in 1755 destroyed the city (mainly the Baixa), which the Marquès de Pombal rebuilt with a rectangular pattern.
2. In the case of Paris, the Romans founded a new city on the higher grounds in the South and built a road going from South to North, which defined an axis on the island: West (administrative)–East (religious), which persisted. In the seventeenth century, the city walls were replaced by broad roads (boulevards), and in the nineteenth century, Haussmann (1809–1891) destroyed many medieval quarters and modernized their structures.
3. After its Christianization in medieval times, Aarhus developed similar patterns in its center as Bremen. The original town hall was constructed with its back to the cathedral's facade and a central marketplace opened in front of it. The town hall was later destroyed and moved twice to the periphery of the historic center. In this respect, Bremen preserved its tradition (probably due to its political autonomy; Aarhus fell under the rule of the Danish kings who resided in Copenhagen). In contrast to Bremen, the harbor in Aarhus is still near the city center.

Urban morphogenesis generally unfolds in a permanent process in which the geographical and geological conditions are relatively stable. However, the characteristics of a harbor town may be lost if the sea-level changes or new demands invalidate former affordances. Thus Paris is no more a harbor for ocean ships, Bremen had to transform its river into a deep channel in the late nineteenth century to remain a harbor city, but container ships prefer to land in Bremerhaven. Aarhus still has a harbor near the city center, but the Baltic Sea has no tide, and ocean ships prefer

Copenhagen or ports on the ocean side of Denmark. Finally, as in most classical harbor towns, the harbor has moved away from the city center in Lisbon.

4.7 Final Remarks: Visual Morphogenesis, Its Central Principles, and the Proper Concept of Space

As sources of stability and evolution/development in visual art, four significant factors come into consideration:

1. Principles of visual perception, memory, and cognitive processing in humans.
2. Objective properties and design options. They are mainly based on the physics and chemistry of the materials. This factor is particularly evident in the means of painting, the structure of stone and wood in sculpture, the techniques used to print graphics, and the casting of bronze sculptures. In architecture, these are technical innovations such as steel construction or concrete structures.
3. The social field of art. It concerns values in the political area and the condition of the market. Since the art market is now global and the artist has to see himself as a worldwide player, economic/political frames and restrictions are effective. They can generate general trends and standards of value.
4. A fourth determinant is neither cognitive/material nor social/political. Nevertheless, it played an essential role in our analyses. It is the complex of space and time (or space–time) and its role in perception and semiosis. As space and time are central topics of philosophy and humanities, it is worth taking a closer look at this determinant.

Newton's "absolute space", which Kant still used as a foil for his a priori of perception (Anschauung), is challenged. Due to Einstein's theory of relativity and quantum physics, Newton's space is only valid to a limited extent.[14] The physical limitations apply to the vast (outer space) and the very small (the subatomic realm). However, facts of vision and human perception or semiosis occur in the spatial scale's middle range. Euclidean space is central to human perception and locomotion but not universal. Helmholtz already showed in a lecture of 1870 based on Riemann's mathematical results that non-Euclidean (constantly curved) spaces are also relevant in human vision. Nevertheless, he made it clear in his rectorate speech of 1878 that Euclidean geometry represents a sufficiently good approximation for the visual experiences of humans.[15] The criticism by relativists, rejecting "objective" models of

[14] Einstein invested the physical space with geometry and dynamics that explain gravitation. In the same mode, visual space (and, by analogy, the space of the other symbolic forms) can be invested with specific features that explain the basic laws of visual perception and communication. Unfortunately, this suggestion cannot be followed in the present context. For the epistemology of symbolic forms and adequate ontologies, see Chap. 7.

[15] Cf. [11] and [12] Brozzoli and Famè ([3]: 153) summarize the results of neuro visual experiments: "Thus, three main regions of space in respect to the body can be differentiated: a personal space, that is the body itself, principally based on the somatosensory input (proprioceptive and tactile

visual semiosis and language, points to restrictions that can be considered. Still, they do not proscribe a scientific analysis of visual morphogenesis or language. Instead, one must use a more flexible conception of space and time.

Concerning the mathematical notions of space, the advances in topology developed towards the end of the nineteenth century are an excellent approach to visual space. General topology can be given various properties, e.g., those of a metric space, but one can also work with topologically weaker (qualitative) spaces. Regarding movement in space, one can apply differential topology, catastrophe and bifurcation theory, and, more generally, differential calculus. Therefore, a range of available mathematical constructs with the help of which a scientific discourse on questions of space and time (of becoming, happening) can be organized beyond the contexts of natural sciences. Cassirer explained in a lecture in 1931 that the physical and mathematical space belongs to the symbolic form of causal-rational knowledge. A somewhat different concept of space can be assumed for the other symbolic forms, although it is not unrelated to the former. A suitable starting point is provided by Leibniz's suggestions in his Analysis situs (cf. Heuser [13]: 186–189). Contrary to Descartes, Leibniz does not presuppose a general, external coordinate system but rather local relations and movements (transformations) that define geometric objects. As he already indicated in his "Mathesis Universalis", this concept leads to a qualitative analysis of space that can also be used beyond a causal-rational construction of space, cf. the arguments in Cassirer [6].

The orders that arise are specific to the particular symbolic forms, and Cassirer speaks of an "aesthetic space" and a "mythical space".[16] The aesthetic space is determined by the different senses and the materials in which artistic forms are realized. This specific nature distinguishes the types of art (painting, poetry, music). Cassirer calls this aspect with Husserl "hyletic" (sensual). This space can be expanded or elaborated and enriched in many ways. Painting, sculpture, theater/opera/film can be combined differently within the visual domain. Music, art, and language can be integrated to form new complexes so that a "total work of art" ("Gesamtkunstwerk" in terms of Richard Wagner) is created.

The space of visual communication has a basis in the primary space of human experience and action, but it is expandable, compressible, and transferable. Social space plays a vital role in architecture and urban planning. This unfolding also applies to aesthetic space as it is subject to social evaluation and is formed in social contexts of use. Since visual perception is essentially dynamic and relates to events and actions, a static concept of space is insufficient. Essential are scenarios of evolution

---

modality, but also important is the vision, which contributes to the definition of this region of space), a peripersonal space region, that is the area around the body, where tactile and visual information are massively integrated; finally the extra personal region of space, whose representation is principally based on the "tele-sensory" modalities (vision and audition).

[16] The deities and demons are arranged in mythical systems and, finally, in religions. In some polytheistic religions, this leads to families or even societies of gods, and in monotheistic religions, mythical space is centered on one God (sometimes with angels and saints). Cf. for further discussions Chap. 5.

and development (morphogenesis), changes (transformations), sudden transitions (catastrophes), and their categorical perception.

In the context of the evolutionary dimension of visual morphogenesis, reference was made to the dramatic changes in human habitation and the consequences that these changes had for symbolic behavior and visual art. One trend is consistent: people are increasingly shaping their environment, first in their immediate area of life (caves, temporary dwellings), then in houses, villages, towns, etc. Nevertheless, the objectively comprehensible space is the ideal point of reference, knowing very well that there is a whole universe of social and individual subsidiary effects that may be decisive in specific situations. In the long-term view, they must be secured in a mathematical language for consistency, e.g., geometry, topology, differential calculus, or a new area of mathematics that must be devised for this purpose.

As these brief comments have shown, a concept of space and becoming/change is necessary for all analyses in the humanities, specifically in the sciences of signs (semiotics), media, and symbolic forms. However, the universality of visual morphogenesis and semiosis depends on two conditions:

1. The analysis must start from a general and consistent construct of space and movement in space. In concrete terms, it must have a geometric-topological and dynamic foundation. In addition, morphogenesis and self-organization are required to cope with the complexity of symbolic systems.
2. Questions of evolution and historical development must be formulated, and answers found. This task overcomes the ahistorical view of structuralism, but it must avoid the historicism of linguistics and other disciplines in the nineteenth century.

References

1. Arnheim, R.: Art and Visual Perception: A Psychology of the Creative Eye. University of California Press, Berkeley (1974) (Second Edition: 50th Anniversary Paperback–Illustrated. University Press Group (2004))
2. Brandt, P.A.: Morphologies of Meaning. Aarhus University Press, Aarhus (1995)
3. Brozzoli, C., Farnè, A.: The space representation in brain. In: Bartocci, C., Boi, L.,Sinigaglia, C. (eds): New Trends in Geometry: Their Role in the Natural and Life Sciences, Chapter 7, pp. 136–177. Imperial College Press, London (2011)
4. Bundgaard, P.F.: Toward a cognitive semiotics of the visual artwork—elements of a grammar of intuition. Cogn. Semiot. (Aesthetic Cognition) **5**, 42–65 (2009)
5. Caliandro, St.: Morphodynamics in aesthetics. In: Essays on the Singularity of the Work of Art. Springer, Cham (2017)
6. Cassirer, E.: Mythischer, ästhetischer und theoretischer Raum, reprinted in: Gesammelte Werke. Hamburger Ausgabe (edited by B. Recki). Meiner, Hamburg, vol. 17: Aufsätze und kleinere Schriften (1927–1931), 2004, pp. 411–433 (1931)
7. Cassirer, E.: Language, myth, and art. In: Bohr, J., Hartung, G. (eds.) Mythos, Sprache und Kunst, pp. 159–184 (Nachgelassenen Manuskripte und Texte, vol. 7). Meiner, Hamburg (1942/2011)
8. Descharnes, R., Néret, G.: Salvador Dali 1904–1989. Benedikt Taschen Verlag, Köln (1989)

9. Foucault, M.: Les mots et les choses. Une archéologie des sciences humaines. Gallimard, Paris (1966) (English translation : The Order of Things. An Archeology of the Human Sciences. Routledge, New York, 1989)
10. Gibson, J., J.: The Ecological Approach to Visual Perception. Houghton Mifflin, Boston (1979)
11. Helmholtz, H.V: Über den Ursprung und die Bedeutung der geometrischen Axiome : Lecture in Heidelberg 1870 (new edition by Dörflinger, G.). Universitätsbibliothek Heidelberg, Heidelberg. https://archiv.ub.uni-heidelberg.de/volltextserver/13156/1/geo2.pdf
12. Helmholtz, H.V.: Die Thatsachen in der Wahrnehmung. Lecture 4th August 1878. In: Vorträge und Reden von Hermann von Helmholtz, vol. 2, Braunschweig 1903 (5th edn, pp. 213–247). Berlin (1878) https://www.historische-kommission-muenchen-editionen.de/rektoratsreden/pdf/Berlin_1878_Helmholtz_b.pdf
13. Heuser, M.L. : Die Anfänge der Topologie in Mathematik und Naturwissenschaft. In: Günzel, St. (ed.) Topologie. Zur Raumbeschreibung in den Kultur- und Medienwissenschaften. transcript, Bielefeld, pp. 183–200 (2007)
14. Jodidio, Ph.: Architektur heute, Taschen, Köln (2002)
15. Kress, G., van Leeuwen, Th.: Reading Images: The Grammar of Visual Design (3rd edn, 2020). Routledge, London (1996)
16. Leakey, R.: The Making of Mankind. Dutton, London (1981)
17. Leyton, M.: Symmetry, Causality, Mind. MIT Press, Cambridge (Mass) (1992)
18. Leyton, M.: The Structure of Paintings. Springer, Wien (2006)
19. Marcos, I.: Le sens urbain. La morphogenèse et la sémiotique de Lisbonne. Une analyse catastrophiste urbaine. PhD Thesis, Aarhus (1996)
20. Marcos, I.: Urban Universals. In: Wildgen, W., Brandt P. Aa. (eds.): Semiosis and Catastrophes. René Thom's Semiotic Heritage. Lang, Bern: (101–125) (2010)
21. Meyer, L.: Englische Landschaftsmalerei von der Renaissance bis heute. Terrail, Paris (translation from the French) (1993)
22. Moore, H.: Catalogue of an Exhibition at the Tate Gallery. ed. by David Sylvester. Humphries Co, London (1968)
23. Moore, H.: Catalogue of an Exhibition, ed. by David Michinson. Fondation Pierre Gianadda, Martigny (1969)
24. Moore, H.: Henry Moore. A Shelter Sketchbook. British Museum Publications, London (1988)
25. Nöth, W.: Handbuch der Semiotik. Verlag J. B. Metzler, Stuttgart (2000)
26. Panofsky, E.: Studien zur Ikonologie. Humanistische Themen in der Kunst der Renaissance. DuMont, Köln (1980)
27. Paquet, S.: Le paysage façonné. Les territoires postindustriels, l'art et l'usage, Les Presses de l'Université Laval, Québec (2009)
28. Pedretti, C. (ed): Leonardo da Vinci. Libro di Pittura. Codice Urbinate lat. 1270 nella Biblioteca Apostolica Vaticana (trascrizione critica di Carlo Vecce). Giunti, Florence (1995)
29. Petitot, J.: Physique du sens. De la théorie des singularités aux structures sémio-narratives. Éditions du CNRS, Paris (1992)
30. Petitot, J.: Morphologie et Esthétique. Maisonneuve et Larose, Paris (2004)
31. Velay, Ph.: De Lutèce à Paris. L'île et les deux rives. CNRS éditions, Paris (2000)
32. Verstegen, J.: Arnheim, Gestalt and Art: A Psychological Theory. Springer, Wien (2005)
33. Verstegen, J.: Reforming visual semiotics: the dynamic approach. Semiotica **200**, 31–48 (2014)
34. Wildgen, W.: Process, Image, and Meaning. A Realistic Model of the Meanings of Sentences and Narrative Text. Benjamins, Amsterdam (1994)
35. Wildgen, W.: The Evolution of Human Languages. Scenarios, Principles, and Cultural Dynamics, Benjamins, Amsterdam (2004a)
36. Wildgen, W.: The paleolithic origins of art, its dynamic and topological aspects, and the transition to writing. In: van Bax, H., Wildgen, W. (eds.): Semiotic Evolution and the Dynamics of Culture. Lang, Bern, pp.117–153 (2004b)
37. Wildgen, W.: Éléments narratifs et argumentatifs de l'«Ultime Cène» dans la tradition picturale du XIIe au XXe siècle. In : Caliandro, St., Beyaert A. (eds.): Espaces perçus, territoires imagés en art. L'Harmattan, Paris, pp. 77–97 (2004c)

38. Wildgen, W.: Metarepresentation, self-organization and self-reference in the visual arts. In: Wildgen, W., van Heusden, B. (eds.): Meta-representation, Self-organization and Art, Lang, Bern, pp. 173–199 (2009)
39. Wildgen, W.: Geometry and dynamics in the art of Leonardo da Vinci. Cogn. Semiot. **5**, 1–30 (2010a)
40. Wildgen, W.: Urban Morpho- and Semiogenesis: Analysis of Four Harbour Towns, Congress of the International Association for Visual Semiotics (AISV-IAVS) in Lisboa, 26th September 2011 (2011) http://www.fb10.uni-bremen.de/homepages/wildgen/pdf/Urban_morpho_and_semiogenesis.pdf
41. Wildgen, W.: Visuelle Semiotik. Die Entfaltung des Sichtbaren. Vom Höhlenbild zur modernen Stadt. [Visual Semiotics. The Unfolding of the Visible. From Cave Pictures to modern Towns] Transcript, Bielefeld (2013)
42. Wildgen, W.: Sculpture, diagram, and language in the artwork of Joseph Beuys. In: Bundgaard, P.A., Frederik, S. (eds.) Investigations into the Phenomenology and the Ontology of the Work of Art, Chapter 14, pp. 243–257. Berlin, Springer (2015)

Chapter 5
The Morphogenesis and Self-Organization of Myth and Religion

Abstract In the first step, the domain of religion (including myth and pre- or post-religious symbolic forms) is delimitated. In the second step, the emergence of myth and religion and the morphogenesis of concrete religious structures in some world religions are described. In the third step, major religions' spatial and temporal characteristics are analyzed (concepts of past, future, apocalypse, and spatial itineraries). Some features of religious morphogenesis are modeled in the format of catastrophe theory. The self-organization of elaborated religions (based on texts) is exemplified in the early history of Christianism. The semantics of religious concepts (God, angel, heaven, etc.) are deviant from everyday discourse as their reference (called tanscendent) is endangered by vagueness and ambiguity. This feature leads to the question of truth in religion, religious pluralism, and the phenomenon of "bricolage" (Lévi-Straus) or systemic construction in myth and religion. The final section deals with xenophobia and racist distortions of history in the political mythology of the Nazi regime.

5.1 Introductory Remarks on Myth, Religion, and Quasi-religious Phenomena

Ernst Cassirer treated the symbolic form of myth and religion in the second volume of his "Philosophy of symbolic forms" (1923–1929). Mythical thinking is embedded in a sequence between natural language (first volume) and mathematically organized science (pure symbolic forms; the third volume of the trilogy). In 1942, Cassirer mentioned another sequence or quasi-order: Language, Myth, and Art (cf. [7]). In his last book ([6] published posthumously in 1946), "The Myth of the State", he focused on myth as a politically manipulating technique, masked as science but below scientific standards. He remarked that the German language had been manipulated by political propaganda, so he could no longer fully understand it at the end of the Second World War. Thus, the symbolic forms Cassirer analyzed after 1920 and until 1945

W. Wildgen, *Morphogenesis of Symbolic Forms: Meaning in Music, Art, Religion, and Language*, Lecture Notes in Morphogenesis,
https://doi.org/10.1007/978-3-031-25651-6_5

allow for transitions and indicate an underlying continuum. The following focus will be on religion with a side view on myth. Religion and theology had been in conflict or rivalry with philosophy since the beginning until philosophy seemed (after the French revolution) to have made religion a non-subject. It became a favorite subject of sociology at the end of the nineteenth century and the beginning of the twentieth century; cf. the contributions of Emil Durkheim, Marcel Mauss, and Max Weber (these aspects are treated in Wildgen [44]: 67–77, in German). The ethnography of myth was the central concern of Franz Boas and his school in the United States, and Levi-Strauss joined this tradition when he was in New York as a refugee. In the series of books called "Mythologiques" that Lévi-Strauss published after his return to France, a vast corpus of mythological narratives was analyzed in a hermeneutic, almost literary style. He was therefore named as a member of the Paris Academy, an honor customarily reserved for authors of novels and poems. Specific aspects of his analyses are commented on in the section on space in myth. The focus of this chapter is, however, on the morphogenesis of myth and religion, not on a typology of myths and religions.[1]

The biological morphogenesis of the human predisposition and interest in religion is probably linked to the emergence of human cognition (the mental grasp on past and future events and actions) and human societies. Like human language, it follows from the formation of Homo sapiens in the Paleolithic area between 300,000 and 200,000 BP (BP = years before present). However, predispositions were fulfilled already at the stage of Homo erectus, i.e., since 1,8 million years BP (see my commentaries on the evolution of symbolic forms in Chaps. 3, 4, and 6).

The area of religion first includes what is given today by the abundance of established religious denominations and their practices; in addition, ethnic forms of religion and forms of religion documented from the past that deviate from today's pattern of religious beliefs must be taken into account. These are often referred to as "myths", in their description, one speaks of mythology (of the Egyptians, Greeks, Germans, etc.). Totem cultures and fetishes also play a role in historical considerations. We will give examples as the investigation progresses. Beyond historically proven phenomena of the religious (we subsume the mythical here), one can presume prehistoric, archaic predecessor phenomena. The species boundaries, in particular the evolution of Homo sapiens, form a time limit (about 300,000 years ago). Politically motivated myths, ideologies, and philosophical systems of ethics, which offer themselves as a replacement or as a modernization of traditional religions, form an upper limit (e.g., the twentieth and twenty-first centuries). The fact that, in these cases, there is no longer any mention of a god, creator, or historically documented prophet is not an exclusion criterion. In the diversity of world religions and their variants, there are certain confessions without a personal god, as in Buddhism and Far Eastern religions (see Glasenapp [12]: 209ff.). The forms of religiosity are outlined by the limits mentioned in Table 5.1.

---

[1] Cf. my comments and the reconstruction of Lévi Strauss' "formule du mythe", a quasi-arithmetic formula that summarizes his insights into the phenomenon of myth in Wildgen ([44], in German and [43], in French).

Table 5.1 The area of religion as the central field of a more general phenomenon

Early stages of man: Homo erectus, Homo heidelbergensis, and Homo neanderthalensis manifest religious practices like burials The species Homo sapiens fulfills the conditions for the emergence of religion	Many forms of religion are historically documented or still exist, including the major world religions and their internal variation. It contains religions with a personal God and those that assume more general religious principles or forces. Beliefs without written traditions are based on oral traditions and ritual initiation and are organized differently than those with a canon of sacred texts[2]	Many religions have developed hybrid forms of public and religious power. Modern secular types of government also show quasi-religious traits or create substitute forms of belief or principles of community ethics
Evolution, prehistory	History; specific references to religious forms, practices, and content	Modernity, present time; substitute forms of religion

An overarching, comparative concept of religion only emerges when dealing with many religions or in a competitive situation. This move is happening very late in the history of religions (in modern times and especially in the nineteenth century). Since colonialism and its Europe-centric perspective have dominated the discourse in both contexts, the term "religion" is sometimes referred to as a "European-Christian invention" (see Haußig and Scherer [16]). The self-designation of a religious community, such as Hinduism or Judaism, usually occurs late and in contrast to other communities. As in Hinduism, one's religion may be understood as a universal phenomenon,the other communities are only partial, incomplete realizations or specifications (cf. Haußig [15]). The controversies about the idea of religion manifest a semiotic discourse within communities and in academia that seeks to force existing regulatory definitions into a coherent grid, whereby the distinctions inherent in religions are modified or ignored. We shall not go this route and take as a starting point a core area outlined by well-documented or empirically accessible religions, myths, rituals, and practices. From this field, archaic predecessor forms can be carefully guessed, or manifestations of faith can be addressed that, after the loss of importance of some church organizations, are a kind of functional continuation of religious behavior and beliefs. These forms are referred to as pseudo-religious.

According to Cassirer, religion is a separate "symbolic form" that shares a semiotic basis with language, art, technology, and science. The relationship of religion to its (possible) predecessor or successor forms is illustrated in Table 5.1.

[2] Émile Durkheim argued that the ritual unfolds social elements present in barter trade. In a similar line, Habermas says that the ritual contributes to the confirmation of social dependency and obedience to rules of behavior (cf. Habermas [13] and Viertbauer [33]: 125).

In addition to theology and religious studies, sociology, psychology, and the philosophy of religion have dealt with the phenomenon of religion. The morphogenetic analysis of religion opens a new interdisciplinary field in which religion is analyzed as a natural phenomenon linked to humans' ecology and cognitive and social capacities.

5.2 The Morphogenetic Approach in Religious Studies

One has to distinguish the original genesis of religious forms from the further development of such structures in social and political contexts. From a morphogenetic point of view, the former is paramount and, under fortunate circumstances, offers the possibility of recognizing regularities, i.e., of gaining general knowledge. The latter first has to be freed from a myriad of historically contingent influences, e.g., coincidences of politics and their consequences, so that the intrinsic dynamics of religion become visible. Eventually, and this is what Max Weber was looking for, the laws of the propagation of religious creeds and the coordination of aligned needs can be found. Scientific clarification of the properly *religious* moment is possible only if the non-religious factors are controlled. The social laws are by no means primarily of a religious nature and are therefore only of secondary relevance for our investigation.

In the context of a "morphogenesis" of the religious, one might first think of the founders of religion, such as the legendary Moses and the historically verified figures of Buddha, Jesus, or Mohammed, in this historical sequence, or of figures such as Thomas Aquinas or Luther, who influenced the transformation of Christianism. The same applies to other religions. There are many divisions and sects within or on the fringes of the major religions. Moreover, many theologians, and even religious laypeople, make their image of God or the transcendent powers; this image may change throughout their biography.

In this ocean of differences, the analysis within the framework of morphogenetic semiotics must first focus on the local processes, that is, on the religious individual in a spatially and temporally limited situation. The global coordination of individual morphogenesis in the family, group, village, country, etc., is a problem where secondary functions such as collective emotions, hopes, and fears come to the fore. The institutional solidification of myth and religion eventually led to a canon of rituals and scriptures and the organization of churches, monasteries, parishes, schools, etc. They bring to light what the followers of Saussure have called a "system of differences". From their point of view, one should understand religion as a shared system of beliefs and rules of conduct. The religious creed would be the individual expression or the implementation of the suggestions and demands given by the institutionalized religion. Only the "arbitrary" conventions of a religious community would be relevant. The individual spiritual life, the faith, would either be the quasi-automatic implementation of the system of religion, a partial realization, or even a faulty deviation. Explaining the laws underlying the phenomenon of faith and affecting the emergence and preservation of religion seems pointless from the perspective of structuralism.

The observability of the moment of genesis, the religious Big Bang, is limited. Therefore, an abductive strategy must be pursued, i.e., more general principles also known from other areas are combined with inductive inferences from observations or measurements in the phenomenal domain. Because of the variety of religious forms, inductive conclusions are either statistical or qualitative in the sense of Max Weber, i.e., one tries to work out basic types. The deductive part can, for example, be based on general laws of morphogenesis. Such are known both from inanimate and from animate nature, and they allow for relevant generalizations. Based on the current state of knowledge, these are as follows:

(a) Evolution and emergence laws. They concern, for instance, the origin of species (cf. Darwin), the origin of life (cf. the hypercycles of Eigen and Schuster [10]) or, in detail, the transition to Homo sapiens, the stages of development of human societies, the emergence of human language and visual and musical forms, technology, and others.
(b) Topologically based morphogenetic models. Thom [28] suggested basic types. There have been similar approaches since antiquity, in modern times, by Goethe (primeval plants and archetypal forms of the skeleton of vertebrates). Thom drew the illustrative material from embryogenesis and geology.
(c) Models of self-organization in complex systems. There is a network of subsystems and levels that are interdependent and coordinated with each other. Global and stable system behavior is achieved by emphasizing (selecting) individual components and eliminating smaller, destabilizing forces or subsystems. Other characteristic properties are auto-reference, feedback, and design for the future. Self-similar patterns, a form of periodicity, can arise that are relatively independent of the substrate and the magnitude of the processes. These phenomena look like rule-based "structures" but result from a complex dynamic process that lacks planning and rule-following.

The inductive path starts from local and momentous forms, as can be found in the case of religion using historical figures and relevant contexts. First, scenarios of the emergence of the religious are discussed, and a theory of stages in the spatial and temporal generalization of the primary emergence is developed. Later, the development and transformation of existing religious systems will be described using historically witnessed figures. Two very individual religious founders, Buddha (fifth century BC) and Jesus Christ (turn of the era) are good examples that allow for the analysis of fundamental principles and processes.

5.3 The Emergence of Religion[3]

As a starting point, one must assume a state of non-religion, which, however, already fulfills the requirements for emergence, i.e., rather vague religious orientations. Let us take as a starting position non-human primates or preforms of humans, e.g., the australopithecines (4–3 million years BP). As with other animals, there are principles of self-preservation, securing survival, on the one hand for the individual and on the other hand for the group (in the case of herd animals). In addition, there is the mother's care for the baby and, in many cases, extended care of a group of relatives for the still-dependent offspring. This example shows that the original principle of care allows a quantitative expansion and thus also a structural enrichment, first from the newly born infant to the group of those born one after the other, who are no longer dependent on the mother's breast. The care activity is expanded such that a group of relatives (fathers, aunts, uncles, etc.) takes care of the child. As a result, complexity accumulates, and feedback and future projections arise. In the original societies of hunters and gatherers, there is also (certainly already in Homo erectus, who made stone tools; approx. 1,8 million years ago) distribution of food in the group. This evolution "socializes" the care of the offspring or other relatives and food distribution. This increase in complexity leads to a higher form of social organization and the emergence of a system of values, initially unconsciously but increasingly as a code of conduct that also has consequences in the event of non-compliance. At the latest with the archaic Homo sapiens, supra-regional networks existed with the exchange of rare objects (feathers, attires) and other valuable goods. There was an expansion of care for strangers who integrated into society. Simple forms of altruism, that is, consideration for others or the adequate distribution of goods, do not yet represent a form of religiosity but are on the threshold of emerging ethics. Expanding the social space from mother–child to families, networks of relatives, clans, local societies, etc., created a precondition for religion.

Both the Neanderthals and Homo sapiens practiced funeral rituals. Through these rituals, a temporal extension of the care beyond death is documented. Those who have played an important role in group maintenance and safeguarding are buried or burned. From this care of the dead, it is only one step to a cult of the dead, i.e., the dead who survive in memory are granted an existence after their death. The spatial and temporal extended care leads to the emergence of a behavior pattern that transcends the here and now; this can be seen as the first level of religious morphogenesis. A distinction can be made between two directions:

a. *Temporal projection.* The dead do not need care; however, extending care until they are buried or cremated makes sense. It is only necessary to transfer the existing empathy to the corpse, especially if one is still obliged to this person. This decision presupposes that there is an intuitive accounting of social benefits. The mother nursing her child has automatic mechanisms of care. For the wider family

[3] This section and further sections of this chapter have been published in an earlier version dedicated to the memory of Per Aage Brandt who died in 2021; cf. Wildgen [45].

and in the case of the distribution of food, socially binding rules must replace the missing instincts, i.e., a kind of bookkeeping, a notion of *just* versus *unjust*, *balanced* versus *unbalanced*, emerges. If the span of thanksgiving is extended beyond the corpse's disappearance, the idea of a "life" after death comes forward. A cult of the dead arises if this idea is extended in time. The dead continue to be cared for; possibly, they can even take revenge in the case of a lack of care. In case the row of ancestors is included, such an expansion leads to a myth of the group's origin. The development of a symbolically coded narrative of the group's origin may be extended to humanity. However, this requires a level of narrative performance, which was only given to Homo sapiens fully equipped with language.

b. *Spatial projection*. We know that primates migrated between neighboring groups. As genetic analyses show, this mainly affected the female members of the groups. They could quickly join an adjacent group. The phenomenon of exogamy, historically verifiable for many human cultures, points to the archaic character of this migration. In addition, rapid climatic fluctuations have regularly squeezed human populations together in narrow spaces. As a result, long-separated and genetically isolated groups have mixed and exchanged cultural achievements. The cooperative relationships between family members and members of a horde were expanded to strangers and thus tended to become a usual strategy. The cycles of significant environmental changes combined with the densification of human populations in ecological "survival niches" turned the spontaneous and temporary adaptations into a characteristic of the species. This kind of ethical globalization is a precondition of religious systems. It does not follow that all populations in all geographical areas have developed structures similar to religions. There may be conditions that have prevented this, e.g., extreme, hostile social fragmentation or an anomic social organization (according to Durkheim [9]), which abolishes existing or emerging social bonds and blocks a vision of that which is beyond ad hoc ethical behavior.

These major emergence lines created an enriched topology of social space and changed the dynamics of social life.[4] We can distinguish two axes:

(1) The *spatial* axis. Values concerning symbolic objects (jewelry, art) and technical innovations were applied to larger exchange areas. Extensive networks emerged between ecologically separated ethnic groups. These were autonomous in everyday life but networked in symbolic and technical areas.
(2) The *temporal* axis. Genetic and cultural "memory" expanded the temporal scope of ethical and religious values. However, the prerequisite for this was not only the cognitive development of episodic and procedural memory but, above all, the existence of a collective memory medium. It mainly presupposes sign structures, most clearly manifested in a common language. Language connects generations,

[4] The concept of "topology of the habitat" was introduced by Lewin [20]. A topology is a local space in the neighborhood of a point. Overlapping local topologies can organize into a more or less coherent map, a global topology. Geographical maps demonstrate the operation at work; in the case of social topologies, constructing a global topology is difficult, as current political crises show.

from infants to the eldest, and generations in the past and future. However, the temporal depth of language only became fully conscious with the invention of writing and thus the permanent fixation of speech.

A first level in this expansion was reached when the caring perspective was expanded beyond the infant and the mother's affectionate behavior to include caring for all children of a woman or a group of relatives. The increase in group size and the extension of the range of locomotion (cf. Dunbar [8]) resulted in the expansion of the behavioral space. This stage should have been reached in the transition from LCA ("last common ancestor"), i.e., from the common ancestor of humans and chimpanzees to Homo erectus.

A second level distinguishes, in particular, the social organization of the Neanderthals from that of modern humans (Homo sapiens). For example, the culture of tools, jewelry, and art (e.g., cave painting) are supra-regional in humans. This fact is particularly evident in the Out-of-Africa movement and afterward, i.e., from 100,000 years on, beginning with the migration inside and outside Africa, later in Europe after 50,000 years B.P.[5]

A third level was reached with the first great empires in Mesopotamia, Egypt, and India. It continues to this day.

The temporal and spatial expansion implies a kind of "transcendence" beyond the phenomena of the here and constitutes a prerequisite for religion. Whether this option is implemented or society achieves a practical (economic and technical) expansion of social space depends on contingent conditions. As the examples of Greek philosophy and science, the Italian Renaissance, and Europe after the French Revolution show, the option of religious transcendence can also be reversed, even erased. But it can come into existence anew as a political ideology, scientific belief in progress, etc. In this respect, religion is not a natural necessity, despite the statistical dominance of religious communities. Only the spatial and temporal expansion of the worlds of experience and imagination seems to be a general development trend in the evolution of Homo sapiens. It remains a mystery why the chimpanzee lineage, our closest relatives, did not follow such an itinerary. In the case of the Neanderthals, a tendency in this direction can be observed; in competition with modern humans. They may not have had the time to take a similar path or to run through it with the same success. They survived only as traces in the human genome due to early intermixture.

In terms of symbolic forms, levels one, two, and three change the reference base of possible symbolic systems. At the zero point, only now- and here-states of the ego and the immediate neighborhood are part of the discourse universe. The early crying of the infant desiring food would, for example, be a gesture of level zero.

At level one, indicators of space (location, direction, distance) and time emerge. They point to the context of immediate action. Those sections of an act that have already been carried out (a kind of short-term memory) or the sections of a complex of activities that are still outstanding ask for an enlargement of space beyond the

[5] New finds in Borneo suggest that at least 40,000 years ago (40,000 to 52,000 BP estimated), there were also Paleolithic cultures in Southeast Asia that produced colored wall paintings (e.g., series of handprints and depictions of animals); see Aubert et al. [2].

immediate present, or "at hand" in the sense of Heidegger's "Sein und Zeit" [17].[6] A group of hominids on the hunt communicates by observing the behavior of those involved, using gestures, and interpreting facial expressions (i.e., without language).

From level two, spatial projections beyond the familiar environment and temporal forecasts into the future (time of shortage in food) or the past (the memory of deceased ancestors and their advice) are part of the reference space. The migration beyond the accostomed seasonal routes and the imagination of history and the future go beyond everyday experience. They create a spatial and temporal "transcendence" as the seed for transcendent ideas in religion. This expanded universe of discourse does not have to be present to everyone. Still, experienced hunters, group elders, strangers who can compare different groups and their way of life, medicine men, and shamans can form the crystallization points of an expanded world of discourse. This experience can be conveyed horizontally through language and its symbolic function, e.g., in narratives and descriptions so that all adults ultimately share it.

At level three, specific media are required in addition to a language. The myths, rituals, imagery and chants, ethnomedical practices, and divination (see the oracles of the Greeks or the bird-flight fortune-tellers of the Romans) form a rich cultural canon. In addition, there is writing and the scientific exploration of space (astronomy, geography) and time (history, utopias).

5.4 The Morphogenesis of Religious Forms

The morphogenetic models presented in Thom [28] assume a germ that unfolds according to the singularities of simple but unstable dynamic systems described in catastrophe theory. More complicated cases, which Thom calls "generalized catastrophes" (cf. ibid. 101–107), can lead to fractal patterns or to the self-similar structures of chaotic systems in which the attractors of the dynamics develop divergently and thus lead to unpredictability.

From the start, there is no stable transcendent quantity (no attractor), e.g., no concept of a hidden entity, a spirit, or a God. Instead, such a concept arises through creating or inventing a core meaning with transcendent content. There are several scenarios in the history of religions that follow these processes:

- Connections are assumed or sought in the environment to explain significant phenomena such as birth, death, rescue, and doom. A hidden agent is imagined; this imagined entity is then increasingly subordinated to human motives. Prayers, sacrifices, or magical practices can be used to control this entity. A world of spirits or invisible agents gradually emerges in the so-called "natural religions" or shamanic rituals. As soon as this world is narratively elaborated and embellished,

[6] These conditions are already met by the honeybees' wagging dance, which they perform when they return to the honeycomb. The space extends to the food source and the time ranges from discovering this source to reaching the source together with the other "recruited" bees. However, the evolutionary framework is different in the case of insects.

a polytheistic world of religious entities (gods/goddesses) can appear. The result became visible in Homeric epics as a drama with gods/goddesses, demigods, and heroes, their heroic deeds, and adventures. Other peoples of the Indo-European linguistic family show similar developments, e.g., the Indo-Vedic world of gods or the world of the Germanic deities. One can observe this polytheistic world of gods in the epics and dramas in which they appear and in the visual arts as a cultural achievement with aesthetic qualities and political functions, specifically in sacral or profane architecture.

- For this reason, these religions are not to be reduced to the moment of the transcendent, the divine. On the contrary, they form an overall cultural achievement whose fascination hardly diminished with the disappearance of the Greek world of gods. After all, it was taken so seriously in the Greek Renaissance that Socrates, who did not fight it but only questioned it in his philosophical reflection, was sentenced to death. Aristotle preferred to emigrate so that the Athenians would not murder another philosopher. Athens' Academy subsequently abandoned the tradition of Socrates, Plato, and Aristotle, and other philosophical schools emerged, among which the Stoa was the decisive one for several centuries.
- The particular case of a radical reduction of the polytheistic world of forms is fascinating. In a long process of meditation and concentration on the essentials, Buddha (fifth century BC) finally reaches a state of enlightenment in which a space of light, nirvana, stands for the divine. An intention in the direction of transcendence remained, but all content-related fillings were rejected as arbitrary embellishments. Comparable to Socrates, who lived some centuries later, the chaos of half-truths and imaginary solutions in polytheism was dismantled by Budha; emptiness appeared in their place. Although the effect of Buddhism on the situation in Palestine or Judaism is not very likely, Jesus also performed such a "purification" concerning Judaism of his time; it later became his doom. He cleaned the temple, where the more or less non-believing high priests (Sadducees) did business and accepted that the Romans confirm their office or sell it to them. He cleansed the law-abiding regulatory apparatus of the Pharisees, who accused him of healing the sick on the Sabbath. As Buddha was the founding figure for "religious cleaners", Jesus is one of his successors. This series also includes Savonarola, a fact emphasized by Luther in the Reformation. The reformers themselves: Luther, Calvin, and Zwingli, also belong to this tradition. The successors of Buddha tried to maintain this radical position in the first centuries after his death. When Buddhism became the state religion in India with the Maurya dynasty and primarily through King Ashoka (approx. 324–232 BC), there was a slow but steady turn of 180 degrees. The founder, Buddha, was worshiped, depicted in monumental sculptures, and became an object of worship. In the sense of Thom's morphogenesis, the eradication of the attractor "God", its dissolution in "nirvana", was reversed. The polytheistic attractor, which had been abolished but was still present in the Hindu environment of Buddhism in India, was re-established and served as a receptacle for new and old contents. In the end, Buddhism disappeared from India, where it now forms a religious minority roughly on the scale of Catholicism (in Sri Lanka, a variant transferred from southern India to the island

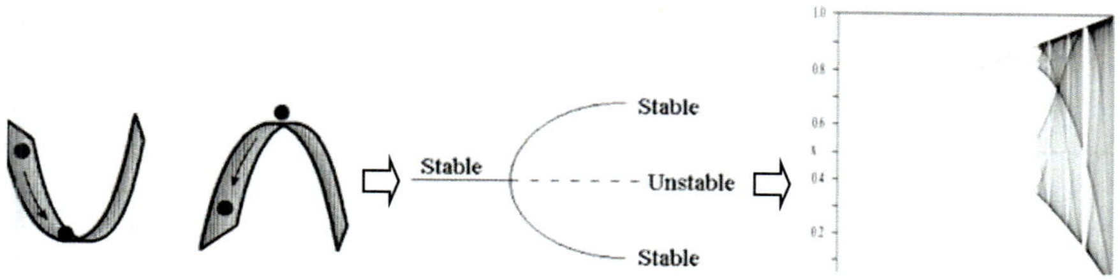

Fig. 5.1 Basic choice attractor/repellor, bifurcation of an attractor, and iterated bifurcations (Feigenbaum diagram)

has remained the majority religion to this day). Nevertheless, the story continued as a northern branch of Buddhism spread to East Asia. Since the religious concepts in Confucianism and Taoism had a similar tendency to Buddhism, insofar as they assumed a rather abstract transcendence of principles, the possibility of integration for Buddhist innovation was more favorable than in India.

The two basic types mentioned above can be schematized with the help of basic notions of catastrophe theory (cf. Wildgen [34, 36]) (Fig. 5.1).

The development may be stopped or frozen in the transition between simple and iterated bifurcations (see the Feigenbaum diagram above). A finite bifurcation sequence leads to a branching or tree pattern, i.e., to a hierarchical system that mimics systems existing in social reality, for instance, the hierarchy of ruling persons in a monarchy or an oligarchy, the order of kinsfolk. Natural morphogenesis in the organic and social world is the template for the morphogenesis of gods/goddesses or ethical principles.

The Greek and Roman heaven of gods/goddesses is a quasi-lexical hierarchy that mimics former monarchic or oligarchic political structures. However, in the polis, the democratically organized society of Athens, such a monarchic organization of heaven, did not match the everyday experience of the citizen. It was Socrates who became aware of this inconsistency. However, the majority in the agora, the forum of democratic decisions, voted for his death, arguing that he was misleading the youth of Athens and acting against religion.

5.4.1 *The Origin of Religious Time and the Role of Memory in Religion*

It is characteristic of many, but not all, religions that time has a beginning in God's act of creation. This beginning is often placed in a historical epoch, e.g., determined based on genealogical tables (see the calculations based on the information in the Bible, for example, in the King James Bible from 1701, which sets October 23, 4004 BC as the first day of creation). The end of time is open; this is God's eternity and heaven's kingdom. Humanity on this earth often ends with a judgment of the world

(Germanic myth) or the Last Judgment (Gospel). The opposition of finiteness (man) and infinity (God) seems fundamental to many religions.[7]

In the lifeworld and human cognition, temporality is closely linked to memory; it exists and is shaped there. Religious experiences and their contexts (times, places, and people) shape individual memories in religious communication. They are crucial when religious content is repudiated or accepted and passed on to new generations. This transmission is vital because it shapes the collective memory and may become a moment of stability in the community's life. It may be called the *past* dimension of religion. Equally important, however, are the instantiation, the application of religious knowledge in new situations, and its further development. This feature can be called the *future* dimension of religious symbols and semiosis. Suppose religion cannot fill the past, especially one's biography, with meaning and provides no useful instructions or perspectives for the future (life, society, world) to follow. In that case, it loses its value and is given up, negated, or even fought with passion. The two axes: past and future, will be treated separately.

5.4.2 The Past Dimension of Religious Signs

Today's religious communities refer to a historical and regional origin, e.g., the Hebrew Bible (Pentateuch) to Abraham, Mesopotamia, and (estimated) the 2nd millennium BC; Buddhism refers to northern India and the sixth century BC. Christianism refers to Jesus (Palestine, first century). In many religions, the historical epoch is uncertain, or reference is made to a mythical early period. Nevertheless, this time and this place are regarded as decisive, i.e., all other events are related to the time of origin, to the time of God's revelation in the case of Abraham, Moses, Jesus, Mohammed, or the "awakening" of Buddha under the Bodhi tree. Although the emphasis is on the immutability of the source experience, the history of religion makes it clear that religion has been decisively expanded, developed, and redesigned. At a historical time (e.g., today), a sign-bound (or text- and image-bound) memory forms the basis of faith and religious practice. Parts of this culture of remembrance can be negated, so the Reformation tried to deny the contribution of the Church Fathers and the papal or conciliar elaboration of Christian teaching. Some religions try to eradicate the remembrance of foreign faiths, i.e., there is a fight for displacement between the religious cultures of memory. Mosaic Judaism erased the traces of the polytheistic culture, i.e., it destroyed the images of their gods. The Christians destroyed sacred trees in Germania or evidence of ancient American religions after the discovery and colonization of America. Mohammed destroyed the images of gods set up in the Kaa'ba, and the Taliban blew up the Buddha statues of Bamiyan in 2008. Book burnings by the Inquisition and similar incidents worldwide belong to the same line. In a way, this is a kind of magical ritual. By destroying an important

[7] For Paul Tillich, the idea of infinity is a defining element of religion. In many cases, however, this would arbitrarily exclude myths and religions that are based in the finite; cf. Tillich ([31]: 73).

symbol of the religion, one believes in destroying the religion as a whole. Even the Holocaust is based on a magic practice designed to destroy Judaism.

As a complement, one can also speak of a command to remember. This command is evident, for example, in Germany's post-war culture and the memory of the Shoah, the extermination of the Jews in the concentration camps of the Nazi regime. The perpetrators and their sympathizers want things to be forgotten, to make the topic taboo in public. The victims and the people who fear a repetition of similar events or wish to prevent them make efforts to set such signs, for example, by memorials either at the site of the event, the former concentration camps, or politically central locations. Figure 5.2 shows in the foreground the monument of the Holocaust in Berlin. It stands close to two other important landmarks, the Brandenburg Gate and the German federal parliament (Bundestag; former Reichstag) with its dome. This parallelism emphasizes the central relevance of the Holocaust memorial because the importance of the other two buildings in its neighborhood is undisputed.

The command to remember must be seen against the background of the automatic processes of forgetting and the disappearance of contemporary witnesses. However, one can neither stop the disappearance of memory nor plaster the urban landscape with memorials. Such memories soon lose their meaning if not kept alive through

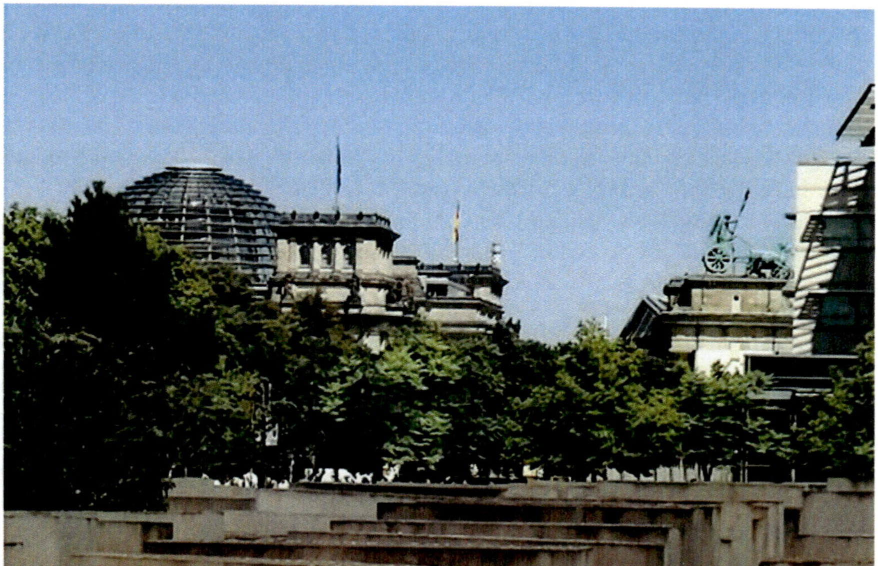

Fig. 5.2 Holocaust memorial in Berlin (in the background are the Bundestag building and the Brandenburg Gate) (Photo, 2022 by WW)

rituals. For example, a museum is attached to the Holocaust memorial, in which this memory can be kept alive through guided tours.

5.4.3 The Future Dimension of Religion

On the one hand, the transcendence of religious symbols means that the afterlife, life after death, and hopes for justice that could not be fulfilled on earth come into focus. On the other hand, the commandments of religion result in expectations and norms that determine the actions of the believer in his future life. This attitude shapes the immediate future. One's biography, the life of the deceased, and that of one's children gain meaning within the framework. A temporal repetition is programmed in the religious ritual. On the one hand, there are sequences of events in time; on the other hand, everything remains the same because the ritual cycle is constantly repeated. Life always returns to the same point; in Judaism, every week for the Sabbath celebration, every year for the holidays: Hanukkah, Purim, Yom Kippur, etc. The week is marked by the Sabbath (Jews), Sunday (Christians), or Friday (Islam); the year is subdivided by festivals (Christmas and Easter in Christianism). In addition, the days may be divided by prayers. This pattern is particularly evident in the monastic rule, e.g., the Benedictine rule. Monastic life is a flow ordered by constant repetition. Even the afterlife can be articulated, for example, in an intermediate zone between heaven and hell (limbo), in purgatory, as a place of transition for sinners who have to serve their sentence before going to heaven.

After the loss of their leading figure, Jesus Christ, the early Christians urgently expected his speedy return, which would still take place in their lifetime and, with it, the Last Judgment and the Kingdom of Heaven for the believers. Underlying this expectation is a turning point, i.e., a sudden, dramatic change that will reshuffle all the cards; all conditions that have hitherto been valid were considered overthrown. Tillich [30] introduced the term "Kairos" (in contrast to "Chronos") for this expectation. The neutral time (Chronos) is interrupted and articulated by a significant moment or phase of upheaval. After two centuries, this belief in an imminent solution to all tribulations was no longer tenable. It was now assumed that the dead would have to remain in a waiting position until the Last Judgment.

Some political ideologies have developed a crass vision of the future. Hitler expected a thousand-year Reich; Marxism adopted historical iron laws through which the final state, communism, must necessarily come about. Here, too, results were repeatedly frustrating. Nevertheless, even if all expectations are disappointing, believing in a specific future is the fundamental force of all religions and ideologies.

5.4.4 *The Dynamics of the End Times (Apocalypse) and Religious Anamnesis*

The historical experiences of the decline of great empires and the occurrence of ecological catastrophes are extrapolated to the future. The signs of decay, death, and destruction herald the end of the world. Such ideas are found in many ancient cultures; they are believed to have derived from Chaldean and Zoroastrian religions, spreading to Hellenistic myths and eventually to Judaism and Christianity. In the Indo-European myths, we find the end of the world, the world conflagration in the Ragnarök of Germanic mythology, and the Pralaya of the Indian Mahabharata. In the Greek saga of Orpheus in the underworld, the world of destruction, of hell, is described in detail and later elaborated in a literary and topographical form in Dante's Inferno. In many cases, there is a cyclical vision of the world's end. For example, a short cycle appears in the Deluge narrative; Abraham concludes a contract with God in which God promises not to deliver the world to another deluge.

The local event of beginning/end or birth/death can be expanded globally via a "pars pro toto"-metaphor. Individual biographical or historical experiences of doom are generalized to the world, the cosmos. A general cause is assumed to be the invariant behind birth and death dynamics. For instance, a capricious, vengeful god dissatisfied with his creation appears. Individual suffering and illness are attributed to disobedience or insufficient sacrifice to a demanding god.

Another religious constellation of forces goes back to the Persian dualism of God versus Satan. God has a Janus-shaped, ambivalent character; the negative pole becomes Satan. This tradition was partially adopted in Judaism and Christianity. Satan is a spin-off from the realm of angels or spirits who serve God. He is banished to Hell with some of Heaven's staff, where he establishes his empire. Consequently, the Antichrist can also be conceived as the opposite pole to Christ. Elaborate Antichrist mythologies make him appear as a negative mirror-design of Christ. The passion of Christ is interpreted as the world's end and the beginning of the kingdom of God. In the Gospel, the sun stands still, and darkness spreads when Jesus dies on the cross.[8] The creation of the kingdom of God, i.e., the restoration of a (better) world, corresponds diagrammatically to the resurrection of Christ.

Concrete historical experiences also shape the horizon of expectation of the prophets, who warn of a foreseeable future. In the second century BC, Antiochus IV threatened Judaism, dedicated the temple in Jerusalem to Zeus and Baal, neglected the Sabbath commandment, and forbade circumcision. The prophet Daniel prophesied his fall and the triumph of Israel over him. The rebellion of the Maccabees seemed to confirm Daniel's prophecy, for, under their leadership, a Jewish rule arose that lasted until 63 BC when Pompey conquered Jerusalem.[9] Daniel's apocalyptic writing thus extrapolated an actual conflict situation (a civil war in Israel and Judah) and the hopes for a Jewish kingdom in the future. The Christian Apocalypse of

[8] Cf. Matthew 27:45 and 51–54, and similarly in Luke 23:44–48 and Mark 15:44–48.

[9] One can also understand the prophecy as a call, a program, and an encouragement. The prophecy is finally fulfilled through it, which might not have happened without the prophet's vision.

John leans very closely on Daniel's style and content; in addition, there were other apocalypses in the Hellenistic area at the time of its creation that could serve as inspiration.[10] We are therefore dealing with indirect morphogenesis because the existing apocalyptic texts and a situation of threat, fears, and hopes are used to update and vary existing apocalyptic designs.

Similar reorganizations of existing traditions shape the life and teachings of Jesus. He refuses to be a political leader or a prophet of political rebellion, as Daniel and the Maccabees did two centuries earlier. At feeding the 5000 at the Sea of Galilee, he deceives the expectations of the masses and boards a ship that takes him to the other side of the lake; he practically flees, confronted with the population's expectations. When he is nailed to the cross, along with two other rebels against the Romans, he is mocked by the attached plaque "King of the Jews", which refers ironically to the role he has refused.

Apocalyptic visions found their way into the canon of the New Testament as "Revelation of John". Iconic and diagrammatic references, dynamic opposites, and cyclical decay- and renewal events have their indexical foundation in the natural processes everyone can observe. Through the handing down of written reports, they are projected onto the past and the future. The inclusion of past events in a religion, for instance, the Passion of Christ in Christianism, is called anamnesis. It renders the memory of dramatic moments or significant passages in the history of a religion (cf. the "rites of passage" introduced to ethnology by Arnold van Gennep, 1872–1957). Birth and death are such passages and may be commemorated by rituals; the apocalypse is the generalization of the passage from life to the death of a community or humankind.

Apocalypses are also omnipresent in modern industrial and media societies. Reference has already been made to forming myths in political and ideological contexts. In the last 50 years, there have been many apocalyptic films,[11] f.i.: The Last Shore (1959), Night of the Living Dead (1968), Bladerunner (1982), The Matrix (1999), and Terminator. The Redemption (2009). Released in recent years: 2017: World War Z, 2018: Quiet Place and Bird Box. These developments will continue because politics motivate multiple fears of environmental catastrophes, e.g., nuclear disasters, climate catastrophes, viral epidemics, or a new world war. The media also pounce on every kind of minor catastrophe, whether in the private circle of celebrities or criminal and terrorist activities. Social networks are also full of them. At first glance, all of this has little to do with religion, but these events go beyond the usual horizon of experience, are transcendent, and require explanations. The readily available descriptions

[10] Éliade ([11]: 187) points to the Hellenistic tradition of the apocalypses originating in the Orphic visions of the torments of hell. In addition to the Jewish tradition of the prophets, the revelation of John also refers to the Hellenistic context (the Roman province of Asia).

[11] Some of the films take up novel-like material again. For example, they refer to science fiction literature, which has been rising since the nineteenth century. The successful authors of the nineteenth century were Jules Verne and H.J. Wells. In the twentieth century, science fiction (since around 1937) became the domain of German (e.g., Herbert Franke), American, Russian, Polish (e.g., Stanislaw Lemb), and Japanese authors.

extensively use the myths and conceptual constructs in religions and thus continue the religious traditions.

5.5 Religious Morphogenesis Described with the Techniques of Catastrophe Theory

In his book "Structural stability and morphogenesis", René Thom has linked morphogenesis in biology and pattern formation in nature and culture. As means of generalization, he used the mathematical results of differential topology, specifically the classification theorem of catastrophe theory. Basic notions are the structural instability of dynamic systems beyond elementary quadratic functions, for instance, the parabola expressed in equation $V = x^2$. A pendulum may illustrate this simple dynamical system. In a dense material, the pendulum goes directly to its resting position. The point of rest is the minimum of the parabola. The negative parabola has an unstable extremum, a maximum (compare Fig. 5.1, at the left). These maximally simple process patterns can be elaborated and deepened using the classification theorem of catastrophe theory (cf. Wildgen [34]: 7–18 and Wildgen [37]: 45–56).[12]

5.5.1 The Birth of a Religious Attractor (Hidden Force or God)

In the following, we argue that the morphogenesis of religions follows such basic dynamic patterns, specifically in the period of their instantiation. Later more complicated dynamics appear. In the context of this book, such a dynamic is sketched without going into the mathematical details. In Wildgen [44], formal model sketches are presented, and in Wildgen [34, 36], the underlying epistemology and the mathematical techniques of catastrophe theory are explained.

The fold is the first system with a stable unfolding, i.e., a change process.[13] Figure 5.3 shows the development of the dynamic nucleus (called singularity). Below this diagram, the same state of affairs is represented by vector flows. They go toward the attractor, away from the repellor, or stop at the level point.

This process takes place on many occasions and concerns. Many individual decisions create a space of coexisting nodes (attractors, repellors). The heaven of gods

[12] The classification theorem of catastrophe theory proves that a series of structurally unstable dynamical systems (singularities) allow for a structurally stable unfolding, i.e., stability exists even in processes following an instability. However, many chaotic and aleatory processes exist beyond these stable and recurrent types of processes. For scientific purposes, the primary aim should be to understand the recurrent, stable processes and later expand the analysis to chaotic or even aleatory processes.

[13] It has the potential equation $V = x^3$. One parameter controls its stable unfolding, the variable u in $V_u = x^3 + ux$ (cf. Wildgen [36]: 111–118).

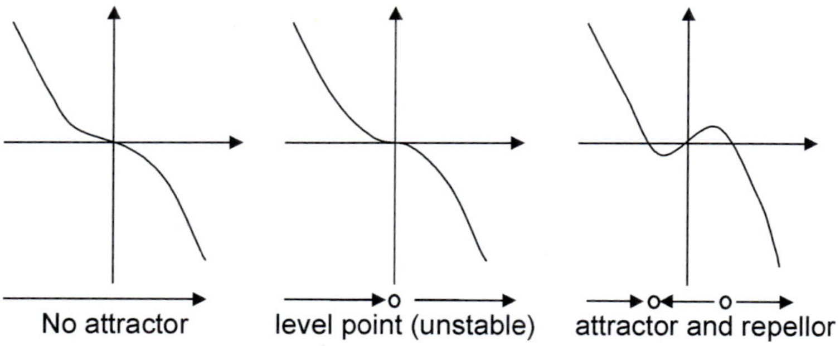

Fig. 5.3 The emergence of a concept of God from an unbalanced system of beliefs

is a consequence of continued morphogenesis that follows this pattern. The original religious need (the "pregnancy" in the sense of Thom [29], cf. Sect. 1.4 of this book) is distributed over many centers (attractors) and thus ultimately reduced in its substance, possibly even nullified. On the other hand, the diversity of religious attractors enables an aesthetic elaboration, which can at least partially compensate for pregnancy loss. An aesthetic, cultural one replaces the biologically rooted pregnancy.

5.5.2 The Implosion of a Polytheistic Religious Universe

The founder of a new religion (e.g., Buddha) can reduce this exuberant diversity to a minimum, ultimately to nothing. The polytheistic multiplicity creates a religious environment where life and the hustle and bustle are omnipresent (the panorama of the Pergamon Altar in Berlin, https://www.asisi.de, is a good illustration). Buddha, on the other hand, meditates in silence under his tree. Buddha's "negative theology" is maintained for a long time by the community of wandering monks until the political success of his religion sets in, and Buddhism puts the charismatic personality of Buddha into the slot of the former God (one of the gods, the most prominent one).

The reintroduction of a divine attractor is not a repetition of the original morphogenesis of many spirits or gods, inferred from a vague recognition of hidden natural forces. Instead, the reduced form continued to be surrounded by a polytheistic culture. The figure of the founder offered itself as an anchor for the reintroduction of a central attractor. The Buddhist doctrine of life remained attached to him such that at least Buddha's mundane and human message was preserved. In the Far Eastern form of Buddhism, there was integration with the ideas of Confucianism and Taoism, whereby parts of the Buddhist heritage found a continuation.[14] The underlying

[14] The history of Chinese religion and philosophy is too complicated to be discussed here. Buddism had a strong influence during the Tang dynasty (618–907) and led to a Chinese variant of Zen Buddhism, which later expanded to Vietnam, Korea, and Japan. The classical positions of Confucianism and Taoism remained points of orientation. Confucianism is still influential indirectly via

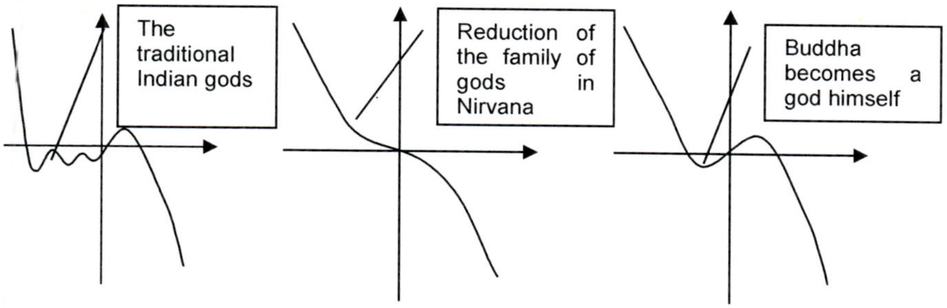

Fig. 5.4 Reduction of a rich landscape of forces and rebirth of a religious attractor

dynamics are much more complicated than the re-emergence of a central attractor in Fig. 5.4 can show.

The reduction is relevant for the overall dynamics of human religions insofar as the original motivation was lost in the diffraction of polytheistic gods and goddesses. The religiosity of polytheism was a burden for the believers they could not shoulder. They became dependent on mediators, priests, scribes, and theologians and were ultimately at their mercy.

5.6 The Self-Organization of a Complex Religious System. The Case of Christianism

In the previous sections, first, very global stages of development and, second, more local morphogenetic structures were discussed. In between are the "systems", i.e., accumulations of religious motifs and the organization of a network of relevant connections. The quantitative increase of mythical and religious motifs and their systematizations lead to self-organization. The drastic reduction (selection) in self-organization secures that only a few parameters determine the qualitative behavior of the system and make it understandable and teachable. Other factors are dampened or eliminated. Self-organization comes forward mainly at branch points, where new solutions appear. Despite its complexity, the system becomes roughly controllable and, therefore, understandable. In the case of religious self-organization, historical, qualitative modeling is the nearby option. In the following, two basic types of religious self-organization are described in more detail:

- The emergence of the main attraction, Jesus Christ, and the construction of his divinity.
- The emergence of the cult of Mary in the early Christian church.

---

its pragmatic attitude centered on governance and public virtues. Mao Zedong made Marxism, Historical Materialism, and thus Atheism the dominant philosophical and religious teaching.

In the Gospels, the person Jesus, his actions, and his speaking are clearly at the center. His life story is described as a movement toward crucifixion and resurrection (ascension). However, after his death, a large void is created first. Christ as a miracle healer and speaker, as a teacher, is no longer in the world. He can no longer be addressed personally or asked for help and advice. Instead, he lives on in the memory of the direct and indirect witnesses.

In the final scene of the crucifixion, only the women "who followed him when he was in Galilee" (Mark 15:40) are among the audience. In the Gospel, women appear repeatedly, but the disciples stand in the foreground. These are now frightened, hide, or even deny that they are acquainted with Jesus, like Peter. One can assume that the painful vacancy of the leader, the leading figure in this situation, is at least emotionally compensated for by the women. The further development of Christianism up to Constantine's Tolerance seems to have essentially been in the hands of committed women. This feature is understandable insofar as the teaching of Jesus focuses on charity and the relationship between man and woman. These contents probably appealed to women, and they still do today. In this reorientation process, a second attractor emerges, which crystallizes in the person of the mother of Jesus, Mary. Our analysis examines the two attractors and shows how content-related elaborations and reinterpretations pile up around them. It is no longer enough to have a nameless god and to strictly follow his complicated set of rules (the Pharisees) or to make enough sacrifices in the temple to satisfy the priesthood (the Sadducees). The religious pregnancy must be concentrated on ideal persons who can be addressed in the imagination and a quasi-dialogical prayer. The Israelite god of law or war could not meet this expectation, particularly pronounced among the ordinary people. Jesus forms the center in the first phase, i.e., as long as he is still alive (or present). Later, a second focus was created, Mary, the mother of Jesus. However, with the addition of these attractors, the original monotheistic God was pushed into a secondary role. The early Church and later the Reformation had to try, if not to solve the difficulties of this unusual and, according to Gestalt laws, unbalanced constellation, at least to avoid evident inconsistencies.

5.6.1 The First Attractor: Jesus Christ

The first question concerns the nature of the force that characterizes the personality of Jesus. It allowed him to persuade a large public (but not the religious authorities) and to attract many followers (the twelve apostles and numerous women who followed him to Jerusalem). Several members of the Jewish elite supported him and got permission to descend his corpse from the cross and place it in a private tomb. These effects point to the characteristic form of Jesus' life, his "Lebensform" in the sense of Wittgenstein. His fundamental decisions were against the role of political messiah, his rejection of the dominant religious tendencies in Jerusalem: the temple cult and the absoluteness of religious rules. He underwent the injustice and violence of the high priests and suffered the indifferent brutality of the Roman authorities (his death). His spiritual message deviated from all contemporary variants of the Jewish religion. It is not easy to guess the background of his originality. Still, the enaction of his life[15] and the contours of his personality, as far as the gospel makes them transparent, were decisive not only for the continuity of Christianism after the disappearance of Jesus. They were necessary as a point of orientation in all reforms of Christianism, i.e., in the required corrections enabling to repair of the losses due to a shift in the direction of dominating contextual forces, internal corruption of the institutions after the tolerance edict of Constantine, and the establishment of a church.

The person Jesus, who is recognized by John the Baptist as a prophet continuing his work, first presents himself as a reformer of Judaism. In conflict with the high priests, who control the temple service (and make a living from it), and the Pharisees, demanding strict adherence to all rules, he tries to convince them with his humane message but ultimately has to bow to violence. In addition, he rejects the expectation of forcing an end to the foreign rule of the Romans as the Messiah. All of this takes place within the framework of the Jewish faith. When it comes to a conflict with the high priests, to whom he is brought before as a prisoner, Jesus says (according to the report of the evangelists) that he is not only God's Son, which can mean many things but is God. This utterance was blasphemy and worth the death penalty for the high priests. At the crucifixion, he is mocked as a usurper by the inscription "King of the Jews".

In the first centuries of Christianity, the question of how being human and being God are related to one another is discussed. It becomes the central point of contention and the cause of the first splits, e.g., the Arians. A question that cannot be answered in retrospect is: Did Jesus take the radical turn from the Jewish prophet and reformer to his self-determination as God alongside the only God of Israel himself (e.g., in the face of death) or did the Christian community under the impression of Resurrection interpret Jesus in this sense?

Mohammed, the founder of Islam (c. 570–632), concedes Jesus has a significant role in the prehistory of Islam but rejects its deification as incompatible with monotheism. For him, Jesus remains one prophet among several. Mohammed was threatened with death in Mecca, as was Jesus in Jerusalem. After his military victories, he established a theocracy under his leadership, which was later expanded under his successors. In the religious tradition, he and his successors became the spiritual anchor of the Islamic faith. Although not deified, Mohammed, as well as the canon of writings that go back to him or are legitimized by him, became the central attractor of Islam.

In late antiquity and before the emergence of Islam, Augustine of Hippo (354–430 AD) suggested the formula of the Trinity as a possible synthesis. To this end, the "Holy Spirit", often represented as a dove or a fire sign over the believer, was introduced as the third magnitude. References to a third power appear in the Acts

[15] Enaction is a term pointing to action in context. Basic contributions to enactivism were made by J.J. Gibson (ecological psychology), theories of the embodied mind and in the context of dynamic system theory.

of the Apostles appended to the Gospel of Luke (Acts 2, 3): "And tongues appeared to them divided as if by fire; and it sat on each of them". The trinity of *Jesus–God the Father–Holy Spirit* now stands for the One God and guarantees the monotheistic character of Christian teaching. With the theological reshaping, Christianism left self-organization and became an argumentatively and rationally established faith. This process dominates the controversies surrounding Christianism to this day. At the same time, the religious doctrine breaks away from simple popular belief and becomes a kind of "secret doctrine" of the specialists.

On the one hand, this duplicity had the consequence that there appeared long-lasting co-evolution of theology, philosophy, and science, in which, however, theology claimed predominance. However, on the other hand, in the fifteenth and sixteenth centuries, the leading function of theology and, thus, of the church was questioned, and the conflict between theology and science became virulent. It persists particularly in the evangelical churches of America.

5.6.2 The Second or Additional Attractor: The Cult of Mary

If one follows the account of the Gospel of Mark, there appear women who meet and speak with Jesus. Among them, there was, to the horror of the Pharisees, even a Samaritan woman from the northern kingdom and an adulteress. It can be assumed that both Jesus and his disciples had female partners since unmarried men were not accepted in Judaism. In addition, according to the Gospels, a relevant female group following from Galilee accompanied Jesus to Jerusalem. Still, it is astonishing that the mother of Jesus, Mary, was able to take a position that made her almost the third person next to the God of Israel and Jesus. The key could be the scene under the cross, when essentially Mary (as well as other women) and the disciple John, whom Jesus recommended to his mother as a quasi-replacement son, stood under the cross. The other disciples stayed in hiding, and even Peter, the rock, was supposed to deny Jesus three times before the first cockcrow. The expansion of the original constellation centered on the attractor Jesus had far-reaching consequences for the attractiveness and vitality of Christianism in the Hellenistic and Roman worlds.

The further development of the cult of Mary is closely related to the dispute over the divine or human nature of Christ. In the Council of Ephesus (431), which has dogmatically fixed Mary's virginity, Mary was called the Theotokos (the person who gave birth to God). She became a mediator between God the Father and Jesus and was closer to popular belief than the more abstract construction of Jesus as simultaneously human and God. In the Proto-Gospel of Jacobus, Mary's birth is announced to her mother, Anna, by an angel (cf. Ceming and Werlitz [7]: 73), i.e., Mary's annunciation is set back one generation. Mary is seen as a human being born thanks to divine providence. The Proto-Gospel, which probably originated in Egypt in the second century AD, is viewed in the Catholic Church as apocryphal and outside the canon. However, it had a significant influence on the Eastern Church. It was also perceived as an authentic narrative in art, for example, in the rock grotto Madonna by Leonardo da Vinci (cf. Wildgen [39]).

In the context of the religions known in the Orient at the time of early Christianity, the tradition of great female deities may also have played a role. These include the Egyptian cult of Isis and, in the Greek context, the goddesses Athena and Artemis; the latter was also referred to as a virgin. Even further in the background, the Babylonian fertility goddess Ishtar or the Phrygian Great Mother of God Cybele may have been the model.

5.7 Religious Topographies. The Bororo Village and Routes of Pilgrimage

Since the symbolic forms and their use depend constitutively on the human body, its movement, and its actions in space, religious signs are also embedded in the area of movements and actions. They form an abstract space derived from the space of movement and action, in which both human actions and imagined worlds have their place. This space can be experienced vividly and most clearly in architecture. If one enters a Christian church, e.g., a baroque or rococo church in Southern Germany, the ceiling often shows a view of heaven populated by angels. The main altar depicts central figures of the history of salvation and prominent figures of saints greeting from the side altars and the walls; the faithful can follow the final phase in the life of Christ, his "Passion". During the mass, the sound of the organ fills the church room, and a polyphonic choir praises God. The heavy church bells and the light bells of the acolytes complete the heavenly impressions. In Rococo churches, the staging of religion gives the impression of a theatrical performance. Religion and theater are close to each other; both contribute to our "spatial memory".

In myth, especially in the archaic forms of totem cultures, there is a much more direct connection between space and myth. Lévi-Strauss described this in detail in his analysis of totem cultures in South America (Lévi-Strauss [19]: Mythologiques). Here the concrete, everyday living space is symbolically subdivided, and this mythical space controls the paths of social action in the village. The spatial organization of the village, its divisions, the living areas, and the respective accessibility for the individuals have their foundation in the myth, which in turn determines the ritual practices. For example, the Bororo village (in Brazil) makes visible the division of a social group into clans, into halves ("moiety") of the dual organization of a village; in the present case, the village community is organized by exogamic principles; see Lévi-Strauss ([19]: 48f.). In the middle is the men's house. The women's shelters form a circle with additional axes and subdivisions. They define the edges of the polygon inscribed in the circle of the village. Married men regularly move between the men's houses and their wives' homes. The axes of this geometric configuration are either defined concerning the movement of the sun or a local geographical entity. In the Bororo village, this is the river that flows along the village (Fig. 5.5).

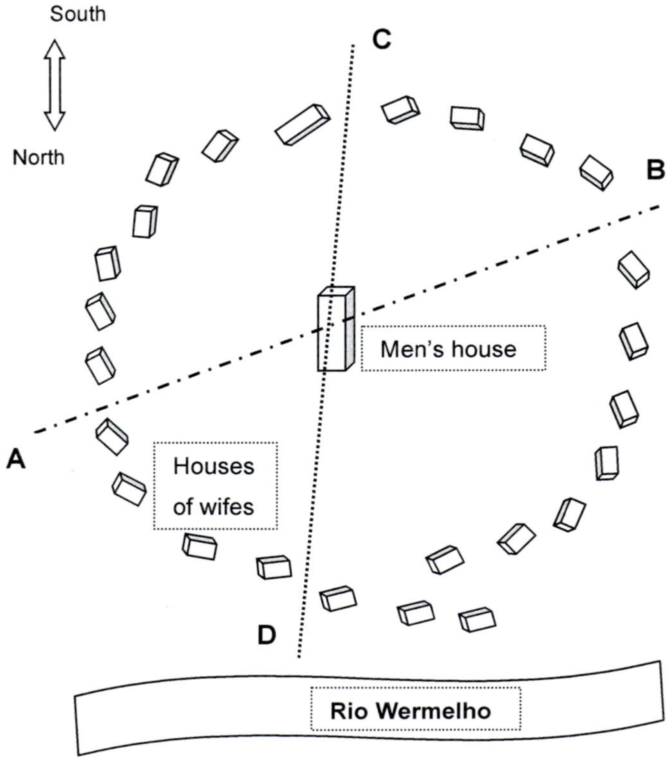

Fig. 5.5 The spatial structure of the Bororo village of Kejara. The line A-B separates the halves and C-D separates the so-called clans. The line C-D is roughly oriented north–south. Line A-B follows the direction of the river; it separates the Cera and Tugaré groups (which also determine the marital networks)

The clans and other social divisions visible in the plan of the village are also marked by "emblems, by privileges [...] finally by ceremonies, songs and proper names" (Lévi-Strauss [19]: 51; translation by W.W.). The myth forms the background of rites and ceremonies. Therefore, we have a symbolic complex that ranges from the spatial organization, daily paths, and routines to rituals and traditional myths. We can say that there is an underlying spatial organization that is very stable in daily practice because of its visibility and its organizing role. This spatial organization reflects the social classifications (e.g., kinship networks) and, finally, the structure and dynamics of myths. In short, the symbolic organization is related to the topology (geometry) of the habitat and the movements in that space. The spatial properties: center, periphery, orientation (global, local), and the dividing axes into two, four, and eight parts visualize the underlying structure and the types of actions appropriate within that framework (see Wildgen [43, 44]; 124–129 for more details).

Religious art is a kind of spatial realization of religion in those faiths that do not know a ban on images. The images in which biblical scenes or scenes from the life of a saint are depicted are like tiny windows into religious imagery. The viewers can compose their private religious world from this lexicon of images. They can create their album with their favorite people, such as Mary or Jesus (e.g., as a child in the crib). When they move around in the everyday world of work, family, or leisure, they can use this world of pictures as a background, as a foundation of meaning. At fixed times, when the bells are ringing, in church, in the cemetery, or during processions in the fields, the religious motion patterns are combined with everyday experience. Especially when everyday life is difficult, oppressive, and uncertain, it becomes bearable through this secondary space rooted in religion.

In those religions that ban images, the visual organization and the imaginary religious space are absent. What takes their place? As a linear sequence of sentences, the written sources do not have an organization comparable to the geometric-geographical spatial structure.[16] However, even in those post-Reformation circles in which the Bible text (in the vernacular) became the central point of reference, believers do not orient themselves to the entirety of the texts of the biblical canon. Instead, the simple layman uses a personal selection of the contents and can choose a few central topoi to construct a conceptual space of low dimensionality.

As early as the Middle Ages, the Catholic Church did a great deal to remedy or at least compensate for the exclusion of believers from the actual ritual performed by the priests. Solemn processions, in which many participants, often arranged according to status and gender, could play their part, passion plays, and scenic depictions of the life of Christ and the saints allowed the population to participate in the religious world of ideas and be actively integrated into the religious world.[17] There are comparable structures in many other religions. This proximity to the believer is the function of the rituals, masquerades, dances, and musical performances with broad participation of the population. Thus, the crowds of believers who travel to Mecca move through the holy places in a fixed ritual and circle the central sanctuary. The pilgrimage movements sanctify a global network of paths (see the pilgrimage to Santiago de Compostela). One can map the whole of Western Europe using the pilgrimage routes. Like the network of paths in an insect colony, the pilgrims measure their continent on foot and thus construe a symbolic map roughly parallel to physical geography.[18]

---

[16] After all, the representation is two-dimensional in an abstract sense, since the one-dimensional syntagmatic space of linguistic chain formation is augmented by the second dimension of paradigmatic relationships, the alternatives in the respective choice of lexemes or constructions.

[17] When the small town Oberammergau (Bavaria) made a vow (in 1634), obliging it to organize a passion play every ten years, more than a hundred places were known for their passion plays. However, only the passion play in Oberammergau subsists. The author witnessed the 2022 version together with more than 4000 spectators. The actors are exclusively citizens of Oberammergau; this means that the play is part of the community life. The actions of Christ, his followers, and adversaries are enacted and thus brought into the immediate neighborhood of the community of believers.

[18] A different kind of geographical-symbolic mapping is generated by trade networks (these may even have been the original mappings in the evolution of humankind) and conquests or wars. For

5.8 The Instability and Degradation of Symbolic Forms in Myth and Religion

Aristotle distinguished "generatio" and "corruptio" in his physics. "generatio" is the equivalent of morphogenesis, the creation of forms, and "corruptio" is the contrary term, the degradation, the degeneration, the corruption of forms. In the following, this contrary process is applied to religious symbolic forms. This process has to consider different aspects of the semiotic triad constitutive for symbolic forms.

5.8.1 Processes of Degradation in the Semiotic Triad: Sign Object–Sign Form–Interpretant

The sign-theoretical triad, according to Peirce, mentions *sign form* (called "representamen" by Peirce; the representing form)–*sign object* (the designated entity)–*interpretant* (the thought, content, or inner image; see below).[19] In the case of the religious sign, e.g., the concept of God, the sign object is God (or, in a polytheistic religion, a specific god or a family, a field of gods). The forms of signs ("representamen" in Peirce's terms) can be the name of the god, its descriptive elaboration through properties, the image of the god (as a pictorial representation, a sculpture, a text.), or its presence in a ritual, e.g., represented by masks or an actor, or a musical or gestural staging of the god. The notion of *interpretant* in Peirce's writings has many aspects. At the phenomenological level, in the world of the sign user, the semiotic process (semiosis) implies causes and forces. Suppose God is the entity (object) referred to by the sign form (representamen), e.g., the name of God, his phonetic, visual, or other representation. In that case, the dynamic interpretant of the sign, God, is the world God created and the believer in this world.[20] God is primarily the creator of everything that exists. He is the germ from which the dichotomy of sign object and sign form (e.g., a word) unfolds. This aspect shows up in the beginning statement of the Gospel of John, verses 1 and 3:

> In the beginning was the Word, and the Word was with God, and the Word was God. [...] Through him all things were made; without him, nothing was made that has been made.

However, this notion of God is highly abstract, and if it meets a philosophical or theological demand, it cannot satisfy the typical adherent of a religious creed. The

---

example, in their campaigns, marches and battles, their settlement in colonies, the Roman soldiers mapped the Roman Empire in their minds.

[19] "The sign or *representamen*, is something that stands to somebody for something in some respect or capacity. It addresses somebody, that is, creates in the mind of that person an equivalent sign, or perhaps a more developed sign. That sign which it creates I call the *interpretant* of the first sign. The sign stands for something, its *object*." (C.P. 2.227–9, cited in Bucher [4]: 99). Cf. for the classification of signs, different definitions, and usages of the notions *sign* and *interpretant* Peirce [23]. In Chap. 7 (of this book) the notion of interpretant is specified beyond the proposals of Peirce.

[20] Cf. the different types of interpretant described in Sect. 7.8.

believer prays to God and asks him for a favor. Although a classical dictum says: "*God helps those who help themselves*", most believers who pray to God and attend religious services instead consider God and other transcendent entities to help them in the same fashion as parents and friends are asked for help. They should have a real-world effect in favor of the prayer. This effect corresponds to the *dynamic* level of the interpretant of the sign God.[21]

In the Christian and Jewish Bible/Tanach, man is considered as the (reduced) picture of God, but God can destroy humankind, as in the case of the Flood, because it did not act as his representation in the world should. The parable about the prodigal son (Luke, 15. 11–32) tells us that the son misbehaves, but his father forgives him when he returns and repents his behavior. Thus, the picture may deviate from the object it represents, but these faults can be repaired. In this case, God acts like a human person and not as an abstract force at the universe's origin. This Janus-headed concept of God may be logically inconsistent, but believers accept such inconsistencies because God is not strictly subjected to human logic.

The existential effect of a religious creed becomes obvious in religious conflicts. Suppose individuals' speech, writing, or religious actions contradict a religious community's norms and standards. In that case, they can be excluded, excommunicated, or banned, which can mean starvation and death. In these cases, a causal impact exists between religious creed and human life. In the inquisition process of the Roman church, the heretic should change his mind and revoke his false belief. Eventually, if he does not, his soul must be saved. Even on the stake at Rome in 1600, Giordano Bruno was shown the cross and asked to revoke his heretic teaching (cf. Wildgen [40]:156f). In traditional Islam, converting to another religion may be penalized by death. The death penalty for non-believers or the burning at the stake for heretics is interpreted as the visible effect of God's force and, thus, proof of his existence. In standard cases, the Last Judgment would decide on the penalization of the believer. In the case mentioned above, everybody can experience the effect and be aware of the dynamic interpretant of God. For the non-believer, the effect results from the religious institution (e.g., the committee of the Inquisition and, in the case of Bruno, the final decision by Pope Clemens VIII).

The interpretant (in terms of Peirce, see above) is the lawful, systematic correlation of the two poles: the object (of the sign) and the form of the sign (representamen). This correlation may be stable for some time. Still, in the sequence of sign usages, either by the same individual or in social exchange, i.e., with variable users, who are naturally different in their acquaintance with semiotic regularities, it tends naturally toward destabilization. Therefore, the instability of the interpretant is a natural process. Different routes to instability can be distinguished. Given a sign form (representamen or representing form), e.g., the word *God, Jesus, Mary* in Christianism or *Jupiter, Mars, Venus*, etc., in Roman polytheism, the reference cannot be secured objectively, and for this reason, it can deviate progressively, lose its contours, and be generalized or specialized. This deformation can affect one term of the duality object versus sign

[21] Probably originated in classical Greek drama, it is widely accepted and also found in the Quran (13:11). It was criticized in Christian denominations that put God's grace to the fore.

form or both. Each singular usage of the sign accumulates aberrations and diverges from the starting point.[22] Two cases must be distinguished:

(a) Instability of the reference (object, e.g., the objects God, Jesus, Mary, Jupiter, Venus).
(b) Instability of the sign form (linguistic, visual, haptic, e.g., the names, pictures, sculptures of God, Jesus, Mary, Jupiter, Venus).

In the first case, two directions are possible. First, God (and, in the same way, the other entities) are generalized. Thus, God may be generalized to the creator of the universe, a universal spirit, or simply to Nature (cf. Spinoza's "Deus sive Natura"). The second direction consists of progressive specifications. God adopts more and more the form of a human agent, listens to the prayers, and answers to them or helps the believer. On the other hand, if the believer does not behave correctly or if his faith is insufficient, God can neglect or even punish him.

In the second case, the name or the visual representation can shift, become similar, or even identical with other names, visual representations linked to neighboring deities. In cultural contact, the name may become ambiguous, designating different deities. The lexical field of religious entities gets blurred, traditional distinctions are abolished, or new distinctions are added. The history of religions tells many stories in this respect. The instability of sign forms leads to unpredictable references.

Religious communities have developed strategies to prevent the instability of religious signs. One strategy consists in avoiding names of God, cf. the biblical name of God YHWH (Hebraic יהוה) that was never pronounced overtly, fearing misuse. Statues and pictures of God were also forbidden. This tendency led to iconoclastic controversies in Judaism, Christianism, and Islam.[23] The other strategy consisted of a dogmatic regulation of all sign-based communications about religion. The canon of holy texts in the three "text-religions" follows this strategy. In parallel, religious institutions control all pictures, statues, and dramatic representations (if not forbidden). The permanent threat of instability in religious communication triggers a wealth of dogmatic and even legal measures and sanctions with the hope that a breakdown of religious meanings can be avoided.

If the indexical/referential function and the reference to something existing and controllable cannot be effectively guaranteed, the creeping instability of religious semiosis cannot be prevented. Semiosis tends to become self-referential. A bulk of fictitious signs fills the mind of believers. This instability is not specific to religion. In everyday discourse, literary cultures, and even scientific discourse, such destabilization and loss of referential security are steady companions of human communication and thinking. Therefore, the instability of religious communication does not devaluate religion or argue for an atheistic or agnostic position. Nevertheless, it indicates

[22] In Chap. 7, the general dynamics of stable and unstable semiotic circuits are elaborated in the context of a dynamic model of semiosis called Kepler semiotics.

[23] The topic of iconoclasm in major religions is treated in more detail in Wildgen ([44]: 104–124), chapter: "Word, Picture or Music. Iconoclasm and Communication about God" (chapter title translated from German). It shows that a rigorous iconoclasm is impossible, as it would abolish any communication, even thinking about God.

the limits of any discourse that misses a solid referential control, the objectivity of scientific semiosis. However, even under the rigid control of referential and semiotic objectivity, the risk of instability, degradations, and sliding into the direction of meaninglessness cannot be permanently avoided.

5.8.2 Attributes of God/Gods and the Instability of Constructs with Abstract Attributes and Universal Quantifiers

For the sake of simplicity, take monotheism as the background. In this case, there are one sign object, God, and correspondent sign forms (representamen, representing forms) in different modalities: the name of God, e.g., JHWH in the Jewish tradition, or a picture of God-father or the Trinity in Christianism, and Allah in the prayers of the followers of the prophet Mohamed. The major attracting force is between God and the world. God has created the world, and (in specific theologies) he regularly and spontaneously interferes with the world.[24] The second force is epistemic, i.e., humans participate in God's knowledge of the world. God's dynamic and epistemic/mental effects establish two gradients (of attraction) that link God to the world and the individual mind. Suppose only the first gradient, i.e., God's causing of the world, is in focus. In that case, the position of God may be reduced to that of a first-mover (as in Aristotelean astrophysics) or to the identification of God with Nature (as in Spinoza's "Deus sive Natura"; God is nature). If God is mainly an epistemic/mental force that allows humans to know the world, we arrive at the mystic union with God in reflection and meditation.

If God as a sign object is blended with its sign form, e.g., his name, picture, or sculpture, God can degenerate into a (wooden, golden) idol. It can be looked at and touched (even destroyed by followers of another religion). If the religion is transmitted via holy texts, the name of God may be conserved, but the content is changed if new attributes are added, and others are neglected. The history of religions, their conflicts, and interferences demonstrate the rich field of diversification and reduction, augmentation and subtraction, and explosion and implosion.

The sign object (cf. Peirce's term), the existent but transcendent God, may become an object constructed only in the imagination. It can be given a concrete form, e.g., an artifact of a craftsman who makes the wooden sculpture of the god or, in the biblical context, the "golden calf". The believer no longer turns to the invisible God but to the idol he can look at and touch. If the priests use this shift to their advantage in the rite and the believers do not (want to) perceive the shift, a deception occurs because, as the biblical critics of the idols correctly tell us, the idol is a human artifact. It cannot satisfy the hopes of help and rescue (even immortality in the afterworld). Investing the idol with divine power in complicated rituals seems to be a clever trick in this context because wood, bronze, and gold remain material substances. This reasoning

[24] In the history of philosophy, the "occasionalism" of Nicolas Malebranche (1638–1715) has further developed this position.

is rationally plausible. However, it presupposes that the sign object, in the case of religion God, actually exists.

It can be pretty helpful to construct such a phantasm as an object of religious fervor since it can, for example, have a political and social effect. Such a positive effect can be seen, for example, in the fact that the community of Israelites is held together by faith in the God of Israel, despite the worst trials over the millennia. Since the scriptures, laws, and rituals are preserved as far as possible, the associated faith can be renewed again and again like a phoenix from the ashes. By reading the texts (again and again), an already lost faith can be revived. In Israel, even the biblical Hebrew was revived in Ivrit (????). Despite the expulsion of Jews from Palestine by the Romans, the religious community survived. The Christians persecuted in the Roman Empire could even achieve a dominant position after the tolerance of the Christian faith by emperor Constantin. After all, due to the colonial expansion of Christian states, it became the most extensive international religious community. Islam united the Arab tribes and enabled them to build a vast empire under the star of Allah.

One could deduce from this that the sign object, God, has always been an illusion. Consequently, at least in faith, there would be no truth, but faith would only have a good or bad effect, i.e., the end of religion would justify the means. Even from such a skeptical point of view, the problem of the stability of the religious attractor remains because, without an anchor in the real world, this stability remains unattainable.[25] As with magical practices, there is always the chance to uncover the tricks of the magicians, the priests, etc. In the sixteenth century, the alchemists were able to enrich themselves with their promise to extract gold from other metals. At some point, however, the deception became apparent; the gold-plated bars were mainly made of lead. The alchemist fell out of favor, and he might have spent the rest of his life in the dungeon. Publicly convicted impostors and fraudsters may be lynched, i.e., operating with fraudulent phantasms harbors real, not just imagined, dangers.

Two significant types of variation and instability concern the sign form ("representamen" in terms of Peirce) and the sign object (the represented).

(a) Variation and instability of the sign form. The sign form can vary or change if the object is fixed. The proper name seems to be the ideal of a rigid designator insofar as it designates an individual (if necessary, specified by further identifications).[26] In the case of common nouns and verbs, e.g., those designating

[25] In his novel "1984", George Orwell describes the dispute of O'Brien, a member of the Inner Party, with the main person in the novel, Winston. Here is a short quote: "[Winston] 'In the end they will beat you. Sooner or later they will see you for what you are, and then they will tear you to pieces.' 'Do you see any evidence that this is happening? Or any reason why it should?' 'No, I believe it. I *know* that you will fail. There is something in the universe—I don't know, some spirit, some principle—that you will never overcome.' 'Do you believe in God, Winston?' 'No,' 'Then, what is it this principle that will defeat us?' 'I don't know. The spirit of Man" (Orwell [22]: 282).

[26] Nevertheless, in many communities, a person changes his/her name in typical transition zones of his/her biography, such as becoming an adult and getting married. A ritualized case concerns the transition of a monk or a nun; they choose a new name as soon as they enter the community of a monastery.

Table. 5.2 Attributes of God in Judaism, Christianism, and Islam, cf. https://www.qcc.cuny.edu/socialsciences/ppecorino/phil_of_religion_text/CHAPTER_1_OVERVIEW/Attributes_of_God.htm

All Just, All-Loving, All Perfect, All-Merciful, All Kind, All-Powerful, Omnipotent, All Charitable All-Knowing, Omniscient, All-Forgiving, All Good, All Understanding, Omnipresent

plants, animals, artifacts, movement, action, etc., one can observe a considerable variation of denominations in geographical space (i.e., in different dialects, sociolects).[27] Scientific nomenclature tried to fix these nouns to ease communication in botany, zoology, and medicine. Biologists like Carl von Linée (1707–1778) have imposed a standard language, i.e., Latin for the scientific terminology, and the Aristotelean hierarchical architecture of designations, following the opposition of class and subclass via a specific difference ("differentia specifica"). However, natural languages rarely show this binary, hierarchical architecture. The meaning of common nouns has one (in some cases several) centers, the core meaning, and a continuous field more or less near this center. At the rims of the semantic field, rather vague transitions separate the different fields; the point of transition is often dependent on the context of use. Cf. Wildgen [35] for a catastrophe theoretical model of vagueness.

(b) Variation and instability of the sign object. In the case of concrete nouns, the sign object can, at least in the case of scientific classification, be specified precisely, i.e., the scientist can control the reference in his use of scientific terms. Such control is usually not possible in natural languages. In the case of abstract nouns, this variability (at the limit of randomness) of reference to sign objects is even more dramatic. Central notions in religion and ethics make use of abstract nouns. For example, if God or a specific god in a polytheistic religion is described, abstract nouns appear with a superlative attribute (universal quantifier "all") added; cf. Table 5.2.

The fact that every single abstract attribute refers to a vague field of sign objects means that accumulating such attributes does not reduce the field. On the contrary, the field's vagueness is increased instead, i.e., the outer border of the prototypical field is enlarged, and the prototypes (the central area) can multiply in the case of polysemy. The field has the most substantial gradient in the center and fades until it reaches the frontier line. The frontier is defined linguistically by hedges, like *quasi-, pseudo-, almost, still*, or by the contact with other semantic fields that impose a transition line.

In the case of expressions like All- and Omni- (cf. Table 5.2), the semantic field is expanded without limits. The conjunction of two simple semantic fields (without a quantifier like All) already has the effect of conjoining the zones of vagueness

[27] The almost random variation of concrete nouns throughout the world's languages has motivated Saussure's "arbitraire du signe".

(the corresponding gradients) and thus enlarging the field.[28] If quantifiers are added and if many quantified attributes are combined, the semantic field proliferates and becomes extremely unspecific; in terms of information theory, it loses information. At the limit, it has information zero, i.e., nothing is said about the entity God.[29]

As an example, take the qualification of God as *All-loving.* The verb *love* has prototypical examples: *the bride loves her bridegroom* and *the mother loves her child.* These prototype attitudes are already different insofar as the first may have sexual/erotic features, whereas the second does not. However, suppose someone loves everybody or even everything in the universe. In that case, the difference between the basic prototypes is blurred, and the content of love vanishes with the magnitude of the set of possible love objects.[30]

If the semantic centers refer to different semantic fields, as in the list above, this disparity risks losing any contour or shape; it becomes amorphous.

5.8.3 Ritual Texts. The Meaning of Hymns or Psalms

The semantic analysis above comes to a result that probably does not persuade believers. They will object that religious texts have a different way of signifying than everyday utterances. Their ritual character, the words' sublimity, and poetic mood have a different goal than profane descriptions, statements, and arguments. Their content is characteristically transcendent, aiming at meaning and communication beyond profane affairs. In the context of sacred objects, buildings, and actions, signifying strives for meanings not easily grasped in everyday contexts. The signification may be compared to the music in a formal context, e.g., in a symphony by Beethoven or Bruckner in a concert hall. We can check this hypothesis for the case of hymns and psalms, typical (poetic) religious texts. A significant effect of the hymn or psalm is the attitude described in the song by Josh Groban, an American singer–songwriter (born in 1981, multi-platinum pop artist):

> You raise me up, so I can stand on mountains.
> You raise me up, to walk on stormy seas.
> I am strong, when I am on your shoulders.
> You raise me up to more than I can be.

As an example of religious (poetic) texts, I take Psalm 46 (King James Version).

---

28 Cf. for the problem of vagueness in semantics Ballmer and Pinkal [3] and Wildgen [35].

29 Prototype semantics is the natural model of meaning in the case of visual prototypes and can be evaluated via computational models; cf. Pino et al. [25].

30 Hannah Arendt analyzed the doctrine of love by the philosopher and Doctor of the Church Augustine in her doctoral thesis "Der Liebesbegriff bei Augustinus" (Arendt [1]). The love to God (not of God) is, for Augustine, the basis not only of human love but of any societal behavior cf. for a semiotic discussion of her argument Wildgen ([44]: 48–52, in German).

God is our refuge and strength,
a very present help in trouble.
Therefore we will not fear though the earth gives way,
though the mountains be moved into the heart of the sea,
though its waters roar and foam,
though the mountains tremble at its swelling.

The psalm lyrically presents an extreme danger, a catastrophe where the earth gives way, mountains are moved into the sea, waters swell, and mountains tremble. God is the only refuge and may help in this extreme and terrifying context. Such situations were probably never experienced by the believers who speak or sing this psalm. Still, the text signifies for them the anguish and scare of a situation for which no one is prepared or can find rational means to save his life. The believers imagine a situation where they are helpless and ask for divine help.

The psalm and many traditional ritual texts transport the memory of terrible catastrophes, passed down by many generations and finally codified in the religious text. Their standard horizon is that of apocalyptic situations. The philosopher and theologian Paul Tillich distinguished two notions of time: Chronos, i.e., the everyday phenomenon of temporal sequence, and Kairos, a time which goes for its end, for a downfall, the doomsday when all rules are overthrown, the reign of God begins, and the final judgment of all human actions is done. The realm of meanings and meaningful utterances on behalf of this time differs from the everyday meanings humans are accustomed to. The standard "logic" of real life in the time "Chronos" is given up. It seems impossible to give an account of such irrational "semiotics" in scientific terms; Kepler's semiotics is not helpful under these circumstances (cf. Chap. 7 for this notion). So the question arises if a poet's semiotics, an aesthetic strategy, may help or if it adds just a new myth to older ones.

5.8.4 Personification and Allegory as an Antidote to Semantic Instability in Religious Systems

To avoid the adverse effects of the instability and variability of individual beliefs and the difficulties of an objective discourse about religion, the history of religions manifests specific solutions: The attributes defining religious entities are projected on examples, ideal persons, life stories of prominent individuals, on visual or enacted figures, e.g., in a tragedy, a novel or a mystery play, a sculptural program. These real or imaginary persons stand for abstract attributes. To reach the corresponding virtue, the believers should follow these persons, identify with them, and imitate their doings or sayings. Thus, in Christianism, the believer, and more specifically the priest, monk/nun who wants to be a saint after death, has to adopt the behavior and, if possible, the appearance of Christ, Mary, or some saint he/she has chosen, e.g., the saint from whom his Christian name is derived. The believer in Islam should follow the behavior or sayings of Mohammed and be like Mohammed. In polytheistic religions, the demigods, e.g., Heracles, Achilles, and many others, serve as ideal persons who should be imitated. The Roman emperors, e.g., Nero (37–68 AD), understood themselves as gods (but mortal), i.e., the transcendent entity has flesh and concrete life and power. The problem of extreme vagueness or a loss of control over the sign object seems to be solved by this procedure. However, this solution is not apt to fulfill the hopes of believers in rescue and justice.

Linguistically, one can observe a double semantic shift: First, by using a proper name, a denomination is rigidly attached to persons; the name even controls their destiny ("nomen est omen"). Second, a shift to description, i.e., the use of common nouns, adjectives, nominal phrases, and narratives. Common nouns and, more dramatically, abstract nouns open a world of variation and instability in the association of sign form and sign object. At the linguistic level, ambiguity and vagueness are increased in metaphors, multiple descriptions, and sentential constructions. This variability in the association of object/state of affairs and signs or sign complexes may be reduced by choosing exemplary persons as incarnations of virtues and attributes of divineness or sainthood (approaching the ideal features of God, Christ, Mohammed). The gospel (of Christ and his apostles), the collection of sayings (hadith) or actions (sunna) by Mohammed, and epical texts such as the Homeric poems, the Iliad, and the Odyssey introduce the personalities of God and their prophets, gods/goddesses, and demigods. They describe their appearance, their actions, and their character. The believers can choose the deities or ideal persons they prefer. Still, they must also consider the deities /ideal persons chosen by their enemies and prepare for a challenge that opposes human beings and the gods they have chosen.

The problem of vagueness, instability, and variation is also relevant beyond religion, e.g., in profane fields that complement or replace religious value systems. An example is given by visual and sculptural programs on late Renaissance facades of town halls in Calvinist towns of the Netherlands and Northern Germany (cf. for the town hall of Bremen, Wildgen [41]: 274–281).[31] The visual program contains a lexicon of public virtues (and vices) that have to be taken into account by their citizen and, more precisely, by leading citizens like the elected members of the municipal council. This topography of virtues and vices elaborates the religious code of Christian (especially Calvinist) communities in a political context. An allegorical composition shows at the left the profane representation of power in the female allegorical figure representing the attribute majesty (lat. "maiestas") holding the imperial "apple" and in the lion representing political dominance. They subdue the pope's power with a tiara and sword, his pontifical cross sticking to his buttocks. The allegorical composition of the vice opulence shows a king on his throne holding down luxury, symbolized by a drunken and naked person. (cf. Wildgen [41]: 279 for analyzing the allegorical program of the late Renaissance town hall in Bremen).

[31] A precursor of these allegorical programs in city halls may be the series of allegorical frescoes in the cityhall (Palazzo Publico) in Siena (Italy) designed by Ambrogio Lorenzetti (1285/1299–1348) in 1338/39: Allegories of Good and Bad Government; cf. https://en.wikipedia.org/wiki/The_Allegory_of_Good_and_Bad_Government.

Political governance is considered in republican constitutions as a system of ethical values parallel to or derived from the world of religion. This example shows that the problem of religious morphogenesis extends easily to social and political life. The consequences will be shown in Sect. 5.9.

In the case of religious concepts, this means that the basic notions like God/gods, angels, heaven, or hell, which are construed by an array of attributes, become semantically unstable, and the association of a sign to its object may become random. This instability has the effect that users can make up their idea of the associated object. Still, it also means that the association of the object cannot be controlled objectively. As a result, religious controversies cannot be solved discursively and may lead to violence, e.g., religious wars.

5.8.5 *Truth in Religion, Religious Pluralism, and Historical Change*

Those religions that have survived the millennia attest to their stability, and one has to ask how this comes about. When we look at music, there may be no truth in music. Still, there is a kind of harmony between the needs and the possible perceptions of musical content, e.g., movement impulses, emotional content, and sometimes quasi-pictorial architectures (cf. Chap. 4 and Wildgen [44]: 116). It can remain open whether there is a quasi-object like the beauty or even the truth of music. The sign object "God" or other transcendent, supreme principles and values in Buddhism and Confucianism must meet specific quality criteria to be accepted and practically implemented by humans as a religious sign system. The problem thus shifts from the question of the truth of religion to its "truth-like", consistent structure that is adequate for human thinking and imagining capacities. If religion aims to achieve long-term stability, it must be convincing or tolerable both for the ordinary believer, the addressee of the religion, and for the technological–scientific or the cultural–educational class to which theologians also belong. In the context of a rational, ultimately scientific discussion of religion, there should be a plausible system of assumptions and conclusions for the reflective person so that religion is acceptable, appealing, and stimulating. To a certain extent, religion is a "total work of art" accessible to everyone. It must therefore have aesthetic qualities. Besides, it has to be adaptable because people and the contexts of life change. Such an adaption is excluded when religion is constructed as an ultimate truth. Rapid changes are dangerous for established religions, as the epidemic fragmentation of religious communities, especially in the post-Reformation era, shows.

In contrast to the creeds coming up in the Reformation, the Catholic Church tried to freeze its dogmatic positions. It took almost four centuries to accept the Copernican view of the universe[32] and three centuries to adopt the liturgical innovations of Luther and other reformers (during the Second Vatican council, 1962–1965); the Darwinian view of the evolution of humans is still rejected in many American Christian denominations. In contrast to the conservatism of large religious communities, an adaptation to historical contexts can be seen as a chance for the survival of a religious community since, from the variety of newly emerging forms of belief, a selection of faith better suited to the new forms of society can be made. Like most Western societies, post-secular societies demonstrate the coexistence of many creeds and agnostic and atheistic positions. The constitutional laws demand the acceptance of other creeds of their citizen, i.e., religious liberalism. This demand induces fundamental changes in the self-image of most established religions. Habermas (1981) interprets this process as the reduction of the traditional sacred, the numinous to everyday discourse open to challenge and compromise (cf. the discussion in Viertbauer [33], Chap. 5).

Despite the variability of religious creeds and the forms of religious practices, there remains the necessity to link these symbolic forms to human experience, human needs, and vital values. Religion must be grasped with the means of the human mind (including feelings and will) and find its place in concrete, spatially realized practices. Religion needs a "fundamentum in re", an indexical rooting in reality. The problem of an ontological rooting of religious beliefs is central (cf. for questions of ontology Chap. 7).

5.8.6 Mythical "Bricolage" and Ratiogenetic Construction of Religious Systems

Although knowledge-based processes are a relevant part of religious systems, these are not primarily knowledge systems but have to respond to other needs and conditions of global success and stability. For a human-relevant religion, at least two aspects are crucial:

1. The doctrine of the faith must also satisfy complex needs, not just offer stopgaps in case of panic. It should exercise a positive, peaceful function that fills the whole psychic field of the individual. This achievement can be affirmed despite all the misalignments in the historically grown religions.
2. The doctrine of the faith must, especially in large, structured, and conflict-ridden societies, contribute to a non-violent coexistence that can use people's capacity to solve problems. This feature requires constant adaptation of the belief bases, at least when implemented in everyday practice. Given the rapid pace of change in the globalized world, this is a tough challenge for religions because they must maintain a credible continuity (what was true yesterday should not be wrong tomorrow) and still react creatively to change. Therefore, a delay tactic

[32] Luther also rejected the theory of Copernicus, but Rheticus was sent by Melanchthon, the theological partner of Luther, to Poland and could persuade Copernicus to have his manuscript printed (in Nürnberg 1543).

is probably appropriate because some temporary challenges may become redundant with the following historical change of context. Most established religious communities follow the strategy: stability first, adaptation only if unavoidable.

Religious systems are very complex organizations and may be compared to an opera by Richard Wagner, where musical, literary, visual/architectonic, and dramatic innovations contribute to a balanced and overwhelming effect. In the history of film, one can also notice developments where, beyond the moving pictures, new forms of music designed for the fifth art are devised, and new techniques of quasi-literary storytelling are invented. This progress means that textual, visual, and musical symbolic forms must be coordinated into a coherent and persuading complex.[33] Of course, the individual spectators, or believers in the case of religion, do not necessarily capture all this complexity. They instead select particular features. Still, the holistic impression of plenitude/opulence persuades them that their belief is part of a global, extremely relevant, and even exhaustive system they can share with a large community.

The dynamics of religious communities are beyond the fulfillment of individual needs, ruled by rather global gradients and flows that are, in many cases, inaccessible to human awareness. Here are a few of them:

(a) During the historical development of religions and their communities, there is a constant flow of augmentation and innovation on one side and loss or reduction on the other. We may call these gradients: (positive) *accumulation* and (negative) *reduction*. In general, the religious system, like other global systems of information and values, prefers accumulation to reduction, i.e., it is harder to reduce or simplify the system than to accept augmentation and addition. In the long run, this results in hyper-complexity, a loss of coherence, and increased contradictions. Moreover, transferring such a hyper-complex system from generation to generation becomes difficult and risky. In a non-literary society that lacks a class of priests, teachers, and administrators, a system that is too complex will naturally collapse and be abandoned. On the other hand, highly organized and wealthy societies may afford hyper-complex religions or other complicated systems of rules and laws. In the case of political and social conflict or war with neighboring societies, the burden of hyper-complexity is brought down, a reduction is enforced, or the whole system is abandoned in favor of another religion, e.g., the one imported by a winning neighboring society or culture. In other cases, the losers of the conflict accept the rules of the winner, but they reduce the foreign system of beliefs to its basic schemes. Thus, in the Christianization of Germanic populations, many belief systems, ethical standards, or technics of medicine and everyday habits were conserved alongside a rather superficial adoption of the Christian faith. Similar processes occurred in the Islamic expansion to North Africa, Spain, and India/Indonesia.

(b) A second gradient concerns the geographical and social distribution of religious systems. The underlying space has three dimensions; two dimensions are given by geographical space, i.e., the surface of a country, its distances, and directions.

[33] Cf. the analysis of film music in Wildgen ([42]: 107–112).

A third dimension is social stratification (in the simplest case, high to low). Religious and other value-based systems (political, economic) result in the recursive tiling of the plane or the three-dimensional socio-geographic space. The problem of regular tilings or tessellations in space was already a topic in Greek mathematics in antiquity, i.e., the regular tiling of a plane by polygonal shapes like regular triangles, squares, hexagons, etc. Bruno, Kepler, Leibniz, and Jungius have discussed (before and after 1600) the tiling of the plane with smaller and smaller circles (and of space with spheres).[34] This problem finally led to the consideration of infinitesimals in the work of Leibniz and Newton. Penrose ([24]: 168–181) discussed non-recursive tilings, which prefigure the problem of fractals and modern complexity theory. In chemistry, tilings are manifestations of diffusion patterns; cf. "Voronoi"-patterns. In the case of religion, the expansion patterns are comparable to a chemical reaction with diffusion and obstacles (such as natural forces like mountains or rivers, lakes or seas).

(c) Moreover, an area may be already occupied, which means conflict, replacement/displacement, or superimposition and interference. If we neglect superimposition, space can be fragmented regularly without lacunas, like an agricultural surface subdivided into fields or claims. After the Christianization of Europe, North Africa, and the near East in late antiquity and early Middle-Age, or after the progress of Islam, from 632 A.D. to 750 A.D., the geographical space was reshaped in terms of religion. Islami expansion defined new frontiers in Egypt, the Maghreb area, Spain, and Northern India, later in the former Byzantine reign, Greece, and the Balkans. Some areas, like Palestine, changed hands between Judaism, Christianism, and Islam and again to Judaism with the foundation of Israel. In all these changes, minorities conserved the former religion, which was tolerated by the majority or led to conflict and war. In feudal Europe until the French Revolution, the populations living under the reign of a sovereign had to adopt the sovereign's religion. This principle was called "Cuius regio eius religio" (whose reign (determines) the corresponding religion).

The conquests of Napoleon replaced religion with the creed of the revolutionary spirit and a republican constitution. It abolished the dominance of religion over profane governance and established a secularist regime. In Germany, Jews became regular citizens and could leave their ghettos. Such geographical and social dynamics can also change the political orientation of the religions involved. The Lutherans and other reform creeds replaced the authority of the Pope with obedience to their sovereign. In contrast, the Catholic Church tried to uphold specific rights of the church concerning the sovereign. Thus, pope Pius XI signed the "Treaty of the Vatican" with Mussolini in 1929, and his follower Pius XII, 1933 still Cardinal Secretary of State Eugenio Pacelli, signed a corresponding concordat ("Reichskonkordat") in August 1933 with the German government (vice-chancellor von Papen). This strategy of soft adaptation to governmental power was called "Aggiornamento" (adapting to the day) and was also pursued as a political device in the Second Vatican Council (1962–1965). With this gradient, principles shared in the states where Catholics live,

[34] Cf. The analysis of Giordano Bruno's geometrical constructions in Wildgen ([38]: 87–102).

like democracy, the acceptance of constitutions and state laws, and free choice of religion were recognized.

Moreover, the dialogue with other religions existent in the same territory (mainly protestant, orthodox Christians, Islamic populations, and Jews) was tolerated. The geographical distribution of religions is beyond political processes correlated with the distribution of other value systems, some older than religious creeds. Thus, archaic family systems influence the preference not only of specific religions but also of ideological and political systems, e.g., political parties representing a religious faith or a political ideology, for instance, a socialist/communist party or a conservative/Christian one.[35]

In his anthropology of myths, Claude Lévi-Strauss tried to find the underlying "logic" and even a mathematical formula for myths (cf. Wildgen [43] and [44]: 225–234). In the context of his anthropology, he introduced the term "bricolage" (handicraft work) regarding religious and other value systems. "Bricolage" captures mythical beliefs, partially ordered or given by chance. They often show a patchwork combining insights and specifications at different times by different subpopulations or neighboring cultures. This strategy is also characteristic of many religions. In antiquity, the polytheistic systems dominant in Egypt, Greece, India, and the Roman Empire could absorb almost all religions encountered during political and military expansion. They could always add a new god to the pantheon, mostly in subaltern places. The adherents to the new god could build their temple and practice their rituals.[36]

Religious innovators like the pharaoh Akhenaten (ca. 1300 BC) reduced the rich pantheon of Egyptian gods to one god Aten (the sun). After his death, the polytheistic system was, however, restored. The religion of Moses was also monotheistic in opposition to the contemporary polytheism of Egypt. It was, however, limited to a specific population, the Jewish tribes, in search of the Holy Land (Palestine), and thus had to acknowledge the existence of other gods in neighboring cultures. In antique India of the fifth century BC, Buddha tried to eliminate all false ideas of religion and eventually arrived at the depersonalized view of "nirvana", i.e., gods and all other entities like heaven, or hell was left apart ("deconstructed" in modern terms). Monotheistic religions like Judaism and Christianism were opposed to the additive strategy of polytheistic religions. Under the Roman emperor Cesar, Jews could obtain a degree of tolerance. Still, Christians were persecuted under the emperor Nero and had to organize in secret places (e.g., meet in the catacombs of Rome). The religious innovators, Christ and Mohammed, purified the complexities of the temple religion in Jerusalem (Christ) and the polytheistic creeds of Arabic tribes around Mecca and Medina.

---

[35] Cf. the publications by Emmanuel Todd, in English: Todd and Garrioch [32].

[36] There existed even a worship for the unknown god in Athens and the apostle Paul referred to it in his speech at the Aeropagus in Athens: "as I passed along, and observed the objects of your worship, I found also an altar with this inscription: 'TO AN UNKNOWN GOD'" (Gospel, Acts, 17: 23). I thank Wolfgang Fritz for this information.

However, these reductions were not the final word. Buddhist creed began to adopt elements of the Hindu pantheon in India; Buddha was represented in statues like a god. In Christianism, Jesus Christ was considered not only the messiah but as God himself at the side of his God-father. Eventually, Augustine created the trinity of God-Father, Jesus, and the Holy Spirit. From a rhetorical perspective, such a trinity was considered a kind of unity, such that the monotheistic quality of Christianism could be preserved. Moreover, with the accumulation of Saints during two millennia of Christianism, an extensive list of holy personalities was created, which were "lifted to the altar", such that praying men/women could address them and ask for advice and help.

Many critical persons inside Catholicism (e.g., Giordano Bruno) rejected the accumulation of saints and their pictures in the religious practice of believers. The reformers, Luther and Calvin, and others suppressed this worship. It seems that all religions practiced by a large population have both tendencies: First, expand the list of objects (persons) of worship, adding pictures, statues, and other material signs. Second, reduce this richness or even ban all pictures and sculptures and limit the objects of worship to only one (or none in original Buddhism).

Lévi-Strauss, in his analysis of South American myths, compared the underlying technology to "bricolage", or patchwork in handicrafts, collage in visual art, and gives a negative account of this activity. On the contrary, it can be argued that religions perform a relevant cultural task. They select among existing ethics and social values, diverse forms of life ("Lebensform"; cf. Wittgenstein [46], and historical achievements (positive or negative). Eventually, they establish a basic set of elements that could direct the future of a group, a society, or humankind. This set is naturally incoherent and possibly contains contradictions and antinomies. The cultural task is, therefore, to find a stable construction that can embed all these selected elements. The architecture of a timber-framed farmhouse may serve as a metaphor for religion. Two different materials, timber and bricks, are characteristic. The timber frame defines the skeleton of the house and its statics. The bricks that fill the lacunas protect the house against wind, frost, rain, and intruders. Additionally, the grounding may consist of field stones; the roof may be covered with thatch or reed. In the case of Christianism, at least three different constituents dominate the theological construction: A very general command of charity contained in all major religions, perhaps with a specific profile and force in the words of Christ (not so much in the deeds of Christian churches); the Messianic ideal and the doctrine to be the people chosen by God (inherited from Judaism); eventually the knowhow of governance taught by the administration of the Roman Empire (adapted after the tolerance edict of Constantine). The last ingredient helped the Christian arch-priests in Rome and Constantinople (Byzantium) to build and consolidate the political constructions in Western and Eastern Europe (the self-proclaimed new Roman Empires, e.g., of Charlemagne or the "Holy Roman Empire of the German nation " until 1806). Similar constructions involving basic ethic values, historical and ethnic identities, and political or historical "grafting" are characteristic of all large religious communities that have survived for millennia.

5.9 Post-Religious Ethics and Political Myth

This section is motivated by the last writing of Ernst Cassirer, "The Myth of the State", and his remarks on ethics (cf. Jagersma [18]). Ethics and practical reason are a foundation level of culture for Cassirer, and all symbolic forms contribute to or are manifestations of ethical principles. In its foundational role, ethics are comparable to myth and religion, and Cassirer considers ethics to be the starting level for both the symbolic forms of language (art) and science. The following section deals with some aspects of ethics and political myth in the spirit of Cassirer.

5.9.1 Protagoras Ethics and the Semiogenesis of Xenophobia

The human body is one model according to which people can shape the image of God. If racist demarcations shape the image of the body, religious terms and ideas may also become racist. The background of this connection is the basic corporeality of the reference system, i.e., one's own body is the starting point (zero point) in the construction of the system of symbolic forms (of language, images, music, and myth). The sentence of the ancient sophist Protagoras "Homo mensura", i.e., "man is the measure of all things", applies at least to the contents of the symbolic forms. The general concepts of the world and God are thus relative to individual conscience or the group's self-perception. Suppose this is closed or self-contained, like the group of Israelite patriarchs, the Pilgrim Fathers in New England, and the Puritan settlers in Virginia. In that case, moral ideals, legal concepts, and religious rules are also limited by this worldview. Socrates and Plato rebelled against such a limitation, as did Kant later in his Ethics. From their point of view, man as a species is the measure of all things and, beyond that, nature, of which man is only a part.

The criticism of "Protagoras' ethics" also comes from systems theory. "Synergetics", the science of coordinating subsystems that achieve global stability and high performance (according to Haken [14]), examines the interaction of many subsystems that limits their dynamics and thus enables controllable, stable dynamics of the overall system. Models are physical systems such as the laser or chemical, and biological systems such as stable reaction/diffusion patterns and regular organization in plants and animals. Such systems are called complex because they assemble diverse, often divergent systems into one overarching "complex system". This feature can significantly increase the overall system's effectiveness but also implies a lack of freedom and an increase of coercion for the subsystems that are integrated into the "synergetic" totality. The group, the clan, the subculture, and the society impose rules of behavior, correct living, and adjusted opinions on individuals or sub-groups. The free motivations and judgments of individuals are regulated or even suppressed. The systems theorist Christoph von der Malsburg describes the willingness to accept such paternalism as a "group instinct". The result is usually a hatred of outsiders and a willingness to use violence against strangers. According to von Malsburg, this tendency of humans and human societies is a fundamental biological attitude of humans, which probably has deep evolutionary roots.

> Group violence is an instability of human society that threatens to strike just anywhere. The neurobiological basis for it is what I call here our group instinct. This instinct is essential as social glue to let us collaborate harmoniously inside our group. However, it does not extend to members of other groups against whom we can be incited to commit arbitrary acts of violence. It is important to raise worldwide consciousness that this instinct is in all of us and that in this sense we are all time bombs. The original structure of the instinct is directed at the social environment we experience directly, but is lifted to large scale by social constructs such as nation, race, class, or religion. These social constructs rest on the misconception that the individual mind is capable of detailed understanding of society as a whole. Overcoming this misconception will rob large-scale group violence of its breeding ground. (Malsburg [21]: 37f)

Christoph von der Malsburg sees a dysfunctional tendency in religions and other overarching cultural constructs. In globalization, archaic principles ("instincts" in his biological perspective) are no longer helpful and have become a danger threatening people and their societies. From this, he derives a critique of religion and other collective constructs. They transport an authoritarian tendency that suppresses the freedom of the individual and thus their creativity. He summarizes:

> It is true that in our society another worldview is still alive, although rarely given clear expression. According to it, I as an individual see myself as part of an organism, which has its own higher goals and reasons, which is high above me and which I cannot hope to understand. My role in life is to struggle and find my place in this world, to conform to what is asked of me, to do my duty, and to find fulfillment in being part of that higher organism. We hear such words spoken in church, but it is sadly true that religion itself to a very large extent is maimed by the Protagorean worldview, formulating apodictic truths to be forced on others [...], picturing a personal god in the image of the individual, a god who is responsible for the creation, who is to be argued with in terms of utilitarian values, and who is misused to define groups and incite violence against non-believers. (Malsburg [21]: 35)

The social instinct is initially limited to the social environment but is later extrapolated to the global world, quickly leading to intractable conflicts. This scenario is reminiscent of Hobbes' hypothesis that man is more like a wolf ("homo homini lupus"). A strict social order (preferably a monarchy in his view at the time of Hobbes) was intended to counter this natural trend. In contrast to von Malsburg's argument, Hobbes considers constructs such as nation or monarchy necessary and beneficial. Opposed to the "group instinct" and Protagoras ethic, there is the phenomenon of altruism, which can be found in all human societies. It arose probably from caring for offspring, mutual help, and the social distribution of food.[37] It plays a central role in most religions; see the "love of one's next" in Christianism. Religion seeks a balance between these contrary tendencies.

---

[37] In this context, the mass of contributions to "altruism" cannot be discussed. However, a scale of choices in the organization of living beings between the egoism of solitary individuals and the altruism of those living in family groups, herds, etc., is a plausible assumption. The species Homo sapiens is in a middle position on this scale. This means that the behavior of individuals in this species is ambivalent or unpredictable. Therefore, any social system and even more religion must consider both tendencies and search for a stable compromise.

There have been, and there are, many totalitarian forms of state and government, i.e., forms of governments that do not grant the citizens any rights but oppress them with brutal force. In the following, the forms that occurred in Europe are analyzed, especially after the First World War, using the example of the Nazi regime in Germany. Comparable is the totalitarian regime in Italy by Mussolini (from 1922 to 1943), in Russia by Stalin (from 1922 to 1953), Portugal by Salazar (from 1930 to 1974), and in Spain by Franco (from 1939 to 1975) and similar regimes in Asia and South America. These global forms (one could even call them symbolic forms as they are supported by propaganda in words and pictures) have arisen in the near past, and their morphogenetic analysis is more accessible than the reconstruction of the genesis of traditional religions, whose initial phase happened thousands of years ago.

From a religious point of view, what is new with these global political forms is, first of all, the radical negation of traditional Christian and humanistic values and norms. In Italy and Spain, the catholic church was tamed or subordinated to the fascist state; after an episode of confrontation, Mussolini achieved "Conciliazione" with the Pope; in Spain, the catholic authorities continued their traditional support for the monarchic state by accepting the general Franco is a temporary substitute; in Japan, the quasi-religious leader, the emperor retained his political and religious functions. The Nazi regime of Hitler and the communism of Stalin tried to abolish the role of Christian churches and replace it with racist or atheist ideologies. On the one hand, religions such as Judaism and Christianism are persecuted or suppressed; on the other hand, an attempt is made to create a short-lived synthesis from given pieces of political, philosophical, religious, historical, and biological discourses of the nineteenth and early twentieth centuries.[38] If in their beginnings, these political movements demonstrated features of morphogenesis and self-organization, i.e., the reduction of the set of political alternatives in favor of radical choices, they became in further development dependent on other strong forces, for instance, military logic, economic pressures, and the proliferation of brutality and murder in a situation of war. As a result, the ideological (religious) components were weakened if not abolished.

5.9.2 *The Political Myth as a Fraudulent Construction: Rosenberg's "Mythus"*

At the end of the Second World War (and of his life), Ernst Cassirer wrote the book "The Myth of the State", describing the misuse of myth and mythical concepts in political conflict, specifically targeted at the Nazi regime in Germany. Rosenberg's book "The Myth of the 20th Century. An evaluation of the mental and spiritual

[38] Leninism and later Stalinism had an intellectually better basis in Marxism, which at least was near to the state of the art of economic and political science in the second half of the nineteenth century. However, Stalinism was quickly overwhelmed by nationalistic and imperialistic contents and opened for anti-semitic propaganda.

creative powers of our time" may represent this myth. Rosenberg, one of the earliest followers of Hitler and his chief ideologist, tried to put the rather impressionist ideas of Hitler and his entourage into a more or less consistent text form. Everything that has a name in intellectual history is related to the new ideal, the Nordic–Aryan race: from Achilles to Alexander, Aristophanes, Aristotle, Aeschylus, Augustine to Zarathustra, and Zwingli. Suppose one examines the respective characterization of the individual persons. In that case, one notices that they are strictly classified according to the given scheme: Nordic–Germanic–Aryan (good) versus Oriental–Near Eastern–Syrian–Jewish (bad). Rosenberg's "Mythus" tries constructing a new lexicon of politically relevant concepts. A network of evaluation standards is brutally reduced and violently compressed into very few, if possible bipolar, attractors. In this black-and-white way of thinking, the fanatic supporter and the opportunist are offered a mechanism of judgment that can lead to a mass consensus. These dynamics are illustrated in the bipolar attractor landscape in Fig. 5.6. Mathematically, the bipolar situation appears in the unfolding of the elementary catastrophe called cusp (with the unstable germ: $V = x^4$).

Many properties and images are assigned to these two poles. Table 5.3 cites passages from Rosenberg's book. The quotations are divided into two sections: External appearance—world view/religion/politics. First, the original statements in German are cited, then (after //) a translation (by the author) is given.

The racial theories had already been enriched with biological analogies at the end of the eighteenth century. The biblical monogenesis theory (Adam and Eve) was replaced with polygenesis theories (origin of genetically different types of people) and models of degeneration (the white race as a pure race). In the second half of the nineteenth century, misinterpretations of Darwin (social Darwinism) gave a "scientific" or modernist touch to traditional antisemitism. In this way, under the guise of modern science and in opposition to a church hostile to science, dangerous myths adorned themselves with the prestige of innovation and modernity.

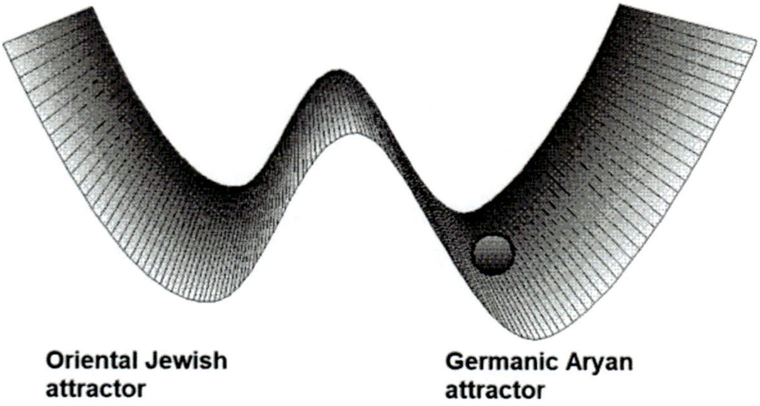

Fig. 5.6 The dynamic concentration of the semantic value space on the two main attractors

Table 5.3 Text passages that illustrate the two semantic poles: Germanic-versus-oriental (in brackets the page in Rosenberg [26])

Äußeres Erscheinungsbild (Outward appearance)	
„hellhäutige Menschen mit goldgelbem Haar "(27); „milchweiße Haut und blonde Haare" //"fair-skinned people with golden yellow hair" (27); "milk-white skin and blond hair"	Die „schwarzbraunen Gestalten" (28), „das fremde, dunkle Blut ... dringt ein" (31) // The "black-brown figures" (28), "the strange, dark blood ... penetrates" (31)
Weltanschauung, Religion (World view/religion/politics)	
(arische Götter) „Heroen des Lichts und des Himmels" (36), „die Anbetung des Lichtes, Weltoffenen, Anschaulichen" (37), (von Winkelmann bis Voß) „Volksabstimmung gleichgerichteter, erdverbundener Charaktere" (56)„ //(Aryan gods) "Heroes of light and heaven" (36) "the worship of the light, the cosmopolitical, the concrete" (37) (from Winkelmann to Voß) "Popular vote of well-directed, earthbound characters" (56)" —— „die einzige noch typenformende Kraft, ..., die Herrschaft des pater familias" (57); „die Ehe" (57); „die Mutter" ... „die edle, fruchtbare Seite des Demeterkults" (47), „der blauäugige, gewaltige Sulla, der rein nordische Kopf des Augustus" (57) // "the only still type-forming power, ..., the rule of the pater familias" (57); "marriage" (57); "the mother" ... "the noble, fertile side of the Demeter cult" (47), "the blue-eyed, mighty Sulla, the purely Nordic head of Augustus" (57)	„Zauberei und niedere Ekstatik" (29), „der den Opferlöffel schwingende und die Opferscheite schichtende Priester" (29), „dogmatisch-philosophische Erkenntnis lockt somit den Instinkt aus seinem Erdreich" (31) „das Triebhafte, Gestaltlose, Dämonische, Geschlechtliche, Ekstatische, Chtonische" (37), „ehrgeizige, abtrünnige Priester wie die Gracchen geben demokratischen Neigungen nach" (57), „Sklavenschutz, Frauenemanzipation, Armenhilfe" (57); „Amazonen- und Hetärentum" (47); „der Ehe untreu" (47); „Gesetz der endlosen Geschlechtsbefriedigung" (47), „Männer ... in weiblicher Kleidung" (47 f.) // "magic and base ecstasy" (29); "the priest swinging the sacrificial spoon and arranging the sacrificial logs" (29) "dogmatic-philosophical knowledge thus lures instinct out of its soil" (31), "the impulsive, formless, demonic, sexual, ecstatic, chthonic" (37), "ambitious, renegade priests like the Gracchi give in to democratic leanings" (57) "Protection of slaves, the emancipation of women, aid to the poor" (57); "Amazonism and Hetaeraism" (47); "unfaithful to marriage" (47); "Law of Endless Sexual Satisfaction" (47), "Men ... in Female Clothing" (47 f.)

The political relevance of the myth constructed by Rosenberg remains doubtful. As Hitler's biography (see Simms [27]) shows, he was a semi-educated, self-taught hate preacher obsessed with very few practical goals. Worldview and philosophical or religious considerations were beyond the intellectual possibilities of Hitler and his prominent companions and therefore played no decisive role in practical decisions.

5.10 Summary and Conclusion

The morphogenetic paradigm in religious science allows an analysis of the belief systems, their internal coherence, and the gradients of change. It can also help to describe the geometry of diffusion, conflict, and the superposition of religious traditions. It furnishes the conceptual and mathematical tools able to describe the stability/instability and the change of religions. The tools of dynamical systems theory can be applied in a qualitative fashion, which allows for the integration of traditional analyses in a descriptive or hermeneutic style. In this sense, the qualitative models proposed on the background of dynamic system theory fulfill the function of a meta-analysis of existing theories of religion and their history. Given the natural science access to the phenomenon of "religion", such an interpretative, qualitative access may seem insufficient. Qualitative analysis can, however, prepare and specify scientific hypotheses that are evaluated using current observations and historical facts. Religious processes are often embedded in long-term processes and thus inaccessible to experiments or observations in empirical field studies.

References

1. Arendt, H.: Der Liebesbegriff bei Augustin. Versuch einer philosophischen Interpretation. Meiner, Hamburg (1929/2018)
2. Aubert, M., Setiawan, P., Oktaviana, A.A., et al.: Palaeolithic cave art in Borneo. Nature **564**, 254–257 (2018). https://doi.org/10.1038/s41586-018-0679-9
3. Ballmer, T.T., Pinkal, M.: Approaching Vagueness. North-Holland, Amsterdam (1973)
4. Bucher, J. (ed.): Philosophical Writings of Peirce. Dover Publications, New York (1955)
5. Cassirer, E.: Language, myth, and art. In: Bohr, J., Hartung, G. (eds.) Mythos, Sprache und Kunst. Meiner, Hamburg, pp. 159–184 (Nachgelassenen Manuskripte und Texte, vol. 7) (1942/2011)
6. Cassirer, E.: The Myth of the State. Yale U.P., New Haven (1946/1974)
7. Ceming, K., Werlitz, J.: Die verbotenen Evangelien Apokryphe Schriften. Marix-Verlag, Wiesbaden (2004)
8. Dunbar, R.: Groups, gossip, and the evolution of language. In: Schmitt, E.A.(eds.) New Aspects of Humans Ethology, pp. 77–89. Plenum Press, New York (1997)
9. Durkheim, E.: Suicide: A Study in Sociology. The Free Press, New York (1897)
10. Eigen, M., Schuster, P.: The Hypercycle: A Principle of Natural Self-organization. Springer, Berlin (1979)
11. Éliade, M.: Histoire des croyances et des idées religieuses, Vol. 2 : De Gautama Bouddha au triomphe du christianisme. Payot, Paris (1978)
12. Glasenapp, H.: Die fünf Weltreligionen. Buddhismus, chinesischer Universismus, Christentum, Islam. Diedrichs, Düsseldorf, Brahmanismus (2005)
13. Habermas, J.: Die akzidentielle Konstitution von Glauben und Wissen. Vol. 1 of: Auch eine Geschhte der Philosophie. Suhrkamp, Berlin (2019)
14. Haken, H.: Synergetics. An Introduction (third revised and augmented edition). Springer, Berlin (1983)
15. Haußig, H.: Zum Religionsverständnis im Hinduismus, Buddhismus, Judentum und Islam. In: Hildebrandt, M., Brocker, M.: Der Begriff der Religion. Interdisziplinäre Perspektiven, pp. 101–111. VS Verlag für Sozialwissenschaften, Wiesbaden (2008)

16. Haußig, H., Scherer, B. (eds): Religion–eine europäisch-christliche Erfindung. Philo, Berlin/Wien (2003)
17. Heidegger, M.: Sein und Zeit. Tübingen. Niemeyer, 19th edition 2006 (1927)
18. Jagersma, A.K.: Der Status der Ethik in der Philosophie der symbolischen Formen. In: Sandkühler, H. J., Pätzold D. et al. (eds.): Kultur und Symbol. Ein Handbuch zur Philosophie Ernst Cassirers, Chapter 13, pp. 276–296. Metzler, Stuttgart (2003)
19. Lévi-Strauss, C.: Mythologiques, vol. I : Le cru et le cuit, Plon, Paris (1964); Mythologiques, vol. II : Du miel aux cendres, Plon, Paris (1967); Mythologiques, vol. III : L'Origine des manières de table, Plon, Paris (1968);. Mythologiques, vol. IV : L'Homme nu. Plon, Paris (1971)
20. Lewin, K. : Principles of Topological Psychology. Harper, New York (1936)
21. Malsburg, Ch. von der: The human group instinct as basis of culture and atrocities. In: Kelso, J.A.S. (ed) Learning to Live Together: Promoting Social Harmony, pp. 31–38. Springer, Cham (2000). https://www.researchgate.net/publication/343722305
22. Orwell, G.: Nineteen Eighty-Four. 2000 Harvill Secker, London (first published in GB 1949) (1987)
23. Peirce, Ch.S.: Collected Papers of Charles Sanders Peirce, 3rd edn. The Belknap Press, Cambridge (Mass.) (1967)
24. Penrose, R.: The Emperor's New Mind. Concerning Computers, Minds, and Laws of Physics. Vintage, London (1990)
25. Pino, O.V., Nascimento, E.R., Campos, M.F.M.: Global semantic description of objects based on prototype theory. Trans. Pattern Anal. Mach. Intell. Preprint June 2019 (2019) https://www.researchgate.net/publication/333679461
26. Rosenberg, A.: Der Mythus des 20. Jahrhunderts. Eine Wertung der seelisch-geistigen Gestaltungskämpfe unserer Zeit. Hoheneichen- Verlag, München (first edition 1930) (1941)
27. Simms, B.: Hitler: Only the World is Enough. Penguin Books, London (2019)
28. Thom, R.: Stabilité structurelle et morphogenèse, Interéditions, Paris (English translation: Structural stability and morphogenesis. Benjamin, Reading, 1975; second, corrected French edition 1977) (1972)
29. Thom, R.: Esquisse d'une sémiophysique: Physique aristotélienne et théorie des catastrophes. Interéditions, Paris [English translation: Semiophysics. A Sketch. Addison-Wesley, Boston 1990] (1988)
30. Tillich, P.: Kairos. 1922. Reprinted in: Tillich, P.: Hauptwerke, vol. 4. Writings in the Philosophy of Religion/Religionsphilosophische Schriften. de Gruyter, Berlin, pp. 53–72 (1922/1988)
31. Tillich, P.: Die religiöse Lage der Gegenwart. Reprinted in: Tillich, P.: Hauptwerke, vol. 5, Writings in Religion/ Religiöse Schriften, pp. 27–97. de Gruyter, Berlin (1926/1988)
32. Todd, E., Garrioch, D.: The Explanation of Ideology: Family Structures and Social Systems. Family, Sexuality and Social Relations in Past Times. Blackwell, Oxford (1985)
33. Viertbauer, K.: Religion und Lebensform. Religiöse Epistemologie im Anschluss an Jürgen Habermas. Pustet, Regensburg (2022)
34. Wildgen, W.: Catastrophe Theoretic Semantics: An Elaboration and Application of René Thom's Theory. Benjamins, Amsterdam (1982)
35. Wildgen, W.: Modeling vagueness in catastrophe theoretic semantics. In: Ballmer and Pinkal (eds.), pp. 317–360 (1983)
36. Wildgen, W.: Archetypensemantik. Grundlagen einer dynamischen Semantik auf der Basis der Katastrophentheorie. Narr, Tübingen (1985)
37. Wildgen, W.: Process, Image, and Meaning: A Realistic Model of the Meanings of Sentences and Narrative Text. Benjamins, Amsterdam (1994)
38. Wildgen, W.: Das kosmische Gedächtnis. Kosmologie, Semiotik und Gedächtnistheorie im Werk von Giordano Bruno. Lang, Frankfurt/Main (1998)
39. Wildgen, W.: Geometry and dynamics in the art of Leonardo da Vinci. Cogn. Semiot. 5, 1–30 (2010)
40. Wildgen, W.: Giordano Bruno. Neun Studien und Dialoge zu einem extremen Denker. LIT-Verlag, Berlin [Giordano Bruno. Nine Studies and Dialogues on behalf of an Extreme Thinker] (2011)

41. Wildgen, W.: Visuelle Semiotik. Die Entfaltung des Sichtbaren. Vom Höhlenbild zur modernen Stadt. [Visual Semiotics. The Unfolding of the Visible. From Cave Pictures to modern Towns]Transcript, Bielefeld (2013)
42. Wildgen, W.: Musiksemiotik: Musikalische Zeichen, Kognition und Sprache [Semiotics of Music. Musical Signs, Cognition, and Language]. Königshausen & Neumann, Würzburg (2018)
43. Wildgen, W.: Morphologie et dynamique des mythes: Une nouvelle lecture de Lévi-Strauss, Conference Actualité de René Thom, Université Paris Diderot, Paris, 15th February2019 (accepted for publication in Estudos Semioticos 2023) (2019) http://www.fb10.uni-bremen.de/homepages/wildgen/pdf/Morphologie_et_dynamique_des_mythes_2019.pdf
44. Wildgen, W.: Mythos und Religion: Semiotik der Transzendenz. Königshausen & Neumann, Würzburg [Myth and Religion: The Semiotics of Transcendence] (2021)
45. Wildgen, W.: The morphogenesis of religion and religious self-organization (with a focus on polytheism and christianism). Cogn. Semiot. **15**(1) (2022) https://doi.org/10.1515/cogsem-2022-2010
46. Wittgenstein, L.: Philosophische Untersuchungen. Philosophical Investigations (Trans. Anscombe, G. E. M., Hacker, P. M. S., Schulte, J.) Revised fourth edition by Hacker, P.M.S., Schulte, J. Blackwell, Oxford (2009) https://edisciplinas.usp.br/pluginfile.php/4294631/mod_resource/content/0/Ludwig%20Wittgenstein%2C%20P.%20M.%20S.%20Hacker%2C%20Joachim%20Schulte.%20Philosophical%20Investigations.%20Wiley.pdf

Chapter 6
The Morphogenesis of Language and Morphodynamic Grammar

Abstract Language is the most prominent symbolic form and has been the focus of classical philosophy and semiotics. This book has placed the main focus on three symbolic forms, music, art, and myth/religion, because embodiment and morphogenesis are better expressed via these forms. The current chapter will complete the picture and view language from the same perspective as music, art, and myth/religion. The first section discusses an array of predispositions for the morphogenesis of human language, a possible intermediate step called protolanguage, and the route toward historically documented languages. The following chapters sketch a morphogenetic view of the grammar of human language with an emphasis on the lexicon and syntax. In this part, the author's research on René Thom's proposals for topological grammar is summarized, actualized, and elaborated.

6.1 Biological Predispositions for the Morphogenesis of Human Language

In the first section, conditions for the morphogenesis of human language capacity are discussed, and the fundamental bifurcations in the evolution of the language capacity common to our species (Homo sapiens) are specified.[1] Preadaptations for language are evolutionary and morphogenetic processes not directly leading to human language. Instead, they are selected for locomotion advantages, the correct detection and categorization of objects and events in the environment, and the stability and the further evolution of social relations (for instance, progeniture, child care, and social cohesion). In the evolution of human language, these factors continue to be relevant and eventually co-evolve with language capacity.

[1] Cf. Wildgen ([74, 75]: Chap. 2: 5–24) for major scenarios of language evolution. Results published after 2004 are summarized in Fitch [25], Dor et al. [20]. The author has also treated some questions concerning the evolution of language after 2004 in Wildgen [77–81] and Wildgen [79]).

W. Wildgen, *Morphogenesis of Symbolic Forms: Meaning in Music, Art, Religion, and Language*, Lecture Notes in Morphogenesis,
https://doi.org/10.1007/978-3-031-25651-6_6

In evolutionary biology, the phenomenon of predisposition or pre-adaptation is known in various species. Cognitive evolution (e.g., of the brain and the sensory organs) probably had a specific advantage in the sensory and motor field. The increased imitative faculties and memory enhanced learning and the establishment of rituals and culture. Walking upright and transforming the forehead and the mouth produced the typical phonetic apparatus of man between the vocal cords and the lips. As language capacity involves motor, sensory, and neural abilities, all three domains must be investigated in terms of pre-adaptation. The development of the larynx is possibly the most specific predisposition.

We distinguish three phases: the predispositions for the morphogenesis of human language, an assumed intermediate stage called protolanguage, and the human language capacity that was decisive for the constitution of the species Homo sapiens.

6.1.1 Motor Programs as Predispositions for the Morphogenesis of Language at an Early Stage

The motor patterns of chewing and breathing could have been sophisticated to develop motor patterns of vocalization. The development of mirror neurons enabling quick learning (copying) of motor patterns from other individuals of the same species would have allowed the quick adaptation to traditions or rituals of vocalization (languages).[2] Possibly a gestured language preceded the syntactic organization and fine motor skills of vocalization and articulation. As soon as the muscular control of hand movements was achieved and learning capacities were increased, partial and ritualized hand movements could support semiolinguistic activities on a gestured basis.[3] Corballis [18] suggests a sequence of evolutionary steps based on the function of mirror cells in the human lineage. He assumes the following steps:

1. Grasping of objects (non-human primates).
2. Facial gestures.
3. Manual gestures (in the miming of events).
4. Conventionalization of gestures (Homo erectus 2 my BP).
5. Phonic language and the capacity of episodic memory.
6. Mental time travel: Who did What to Whom, Where, and Why?

Evolutionary steps corresponding to (1) and (4) to (6) will be the focus of morphogenetic grammar in later sections of this chapter. The difficulties of such a sequence

[2] Lewis [44] reports that Ba Yaka pygmies, hunter-gatherer societies in Central Africa, imitate not only other members of the group but also animal sounds, nature sounds and the sounds of foreigners. Similar conditions probably prevailed in hominin species that lived in forests or at the rim of forests.

[3] Condillac had already considered the hypothesis of a gestural origin of language in the eighteenth century. The plausibility of this hypothesis stems from the parallelism between the gestural communication of deaf-mute persons and the vocal communication of humans without such disabilities. In the twentieth century, Allott [1–3] advocated such a model. The cognitive parallelism of gestures and language was also prominent for McNeill [45].

of scenarios are that transferring motor capacities to language would have created a conflict in using resources (muscular, respiratory, and cerebral). As cerebral resources are very costly, such a transfer is only possible if it "pays". Therefore, one needs a causal/dynamic explanation in which social demands for these changes or other functions involving the survival context or social dynamics are empirically proven (there should exist an immediate payoff). As an alternative or in addition, one can draw on the dynamics of self-organization, enforcement, and rapid selection (cf. hypercycles in the sense of Eigen and Schuster [22]) and others. Evolutionary psychology has followed the first route but remained inconclusive because the payoff is only plausible for specialized, technical contexts, even in current language use. The following sections investigate the second route because it fits into the frame of morphogenesis.

The auditory, visual, olfactory, and tactile senses and motor schemata are necessary for creating stable object concepts and the construction of relations between these. They allow the evolution of a stable semantic framework in natural languages. Gibson ([26]: 46 f.) says:

> Similarly the ability to construct an object image from varied properties is absent among reptiles but present among most mammals. All monkeys and apes construct visual object concepts. Only the most intelligent primates, however, (Cebus monkeys, some baboons and macaques, and all great apes) construct and manipulate relationships between two or more objects. (...) Only humans, for instance, use tools to make tools or construct tools from multiple raw materials and then apply these tools in a second goal-directed object–object manipulation. Humans also by far exceed other primates in their ability to construct objects hierarchically."

The capacity of relational thinking enables complex strategies in the search for food (memory for places, categories of food, time of ripeness, and value for different purposes). It is helpful for the preparation of food (cutting, grinding, cooking) and improves the faculty of collective hunting. Social relations may be better controlled, coalitions and power positions independent from actual force can be managed, and intrigues, strategies, and politics can be devised. Language may have become a primary faculty in the context of this increase in instrumental and social intelligence. The behavioral and social consequences of such a cognitive evolution created the conditions under which linguistic competencies "paid", i.e., they triggered a Darwinian scenario that selected individuals or groups based on linguistic skills. In such a scenario, the (latent) language capacity could have made decisive steps at the time of Homo erectus (ca. 2 million years BP = before the present), allowing the migration of this species to Eurasia.

The growth of the brain is a general survival strategy and represents a trend in the evolution of mammals from basic insectivores upwards. The first massive pressure toward bigger brains occurred during the transition to active daylight hunting in the trees. The major transition occurred when early hominids adapted to life in the Savannah (or the borders of forests) by walking upright and hunting as groups. The cortex and, at a different rate, the brain stem grew most quickly. The temporal lobe and later the frontal lobe increased specifically in the cortex. Linked to the temporal lobes and their growth, the asymmetry between the hemispheres also increased. This feature was not "new" for humans but involved a quantitative change, which may have triggered qualitative and functional changes.

A larger brain involves a set of preconditions and consequences. First, brain size depends mainly on in-utero growth in contrast to the growth of bones and, thus, overall body size, which depends on nutritional and environmental conditions after birth. Second, in-utero growth is controlled by the energy supply available to the mother, e.g., the quality of her food. Therefore, a change in hominid diet was the primary precondition for an increased brain. Third, the individual growth dynamics of the brain and body are another key to language evolution. The birth of the human body in a relatively early stage of brain maturation and the considerable difference in brain plasticity and adaptive capacity compared to chimpanzees constitutes an important pre-adaptation for language use and language learning.

The change in the geometry of the larynx is one of the preconditions for spoken language, and it separates two major concavities; the tongue, which moves between them, can regulate the proportion between these "resonators". This proportion controls the formants, i.e., the central frequency bands of vowels. Thus, the articulation of vowels and velar pharyngeal consonants is due to the deeper and vertically transformable larynx.[4] The reduction of the temporal muscles in humans is another significant feature in the evolution of the phonic repertoire (cf. Fitch [25]: 263). The vertical position of the teeth and the closed circle of teeth in humans make the articulation of frontal consonants (dental, alveolar) possible.

In general, it seems that many anatomical and cognitive dispositions for a spoken language were already present before the separation of the lineages of Neanderthals and modern humans. Calculations based on the mutation rate in mitochondria point to a period of 660,000 ± 140,000 years BP and for the DNA of 440.00 to 270,000 years BP (cf. Dediu and Levinson [19]: 188).[5] It fully evolved in the phase of species formation. Still, it could have remained dormant until dramatic changes in the ecology or migrations out of Africa and inside Africa triggered a quick expansion of linguistic competencies and led to the divergence of languages with the distance of the migration routes.[6] Those who argue that language evolved later, say around 50,000 years BP, must reintroduce some (God-given) miracle or some macro mutation incompatible with evolutionary biology.[7]

---

[4] The nursing baby can still breathe and drink simultaneously because both pathways are independent. A sub-velar position of the epiglottis has also been observed in other primates (cf. Starck [55]: 586).

[5] Based on climatological data and a computer simulation of benign conditions for different species of hominins Timmermann et al. [59] suggest that the evolution of Homo sapiens is located between the disappearance of the Homo heidelbergensis (415,000–310,000 y. BP) and the appearance of Homo sapiens (in the archeological findings) 300,000–200,000 y. BP.

[6] At the genome level, Neanderthals and Denisovan hominids share the variant of the FOXP2 gene that distinguishes humans from chimpanzees. FOXP2 is a gene that was shown to be responsible for disabilities in language development if defective, cf. Fisher [24].

[7] Chomsky [16]: 58) assumes that 100,000 years ago, there were no languages, but about 50,000 years ago, the human language existed in many forms. He writes:"The evidence is compelling that since then the language faculty has remained essentially unchanged" (ibidem). In Hauser et al.

Motor programs and their further morphogenesis are linked to the emergence of stone industries in the history of the human species. The first stone axes were produced around 2 million years BP. They make up the so-called pebble culture. Artifacts not only hint at the cognitive level of humans, but are also linked to social life. To produce artifacts and to keep fire, a socially organized exploitation of the environment, a division of labor, and a mode of the social distribution of products must be in place. This capacity requires rules of collective behavior, and language is the prototype of rule-governed social behavior; it not only helps to represent and enact social behavior, but it is also *the central* symbolic representation of social behavior.

The following section summarizes major bifurcations in the morphogenesis of the language capacity of pre-humans and humans. Bifurcation presupposes periods of structural stability, a transition zone, often dependent on aleatory and multiple causes, and a phase of structural stabilization after the transition. Therefore, adequate models must contain concepts of structural stability and abrupt changes (catastrophes). The mathematical models proposed by René Thom, Hermann Haken, and others are the proper background for such an analysis. We shall, however, limit mathematical arguments to a minimum.

6.1.2 Bifurcation Scenarios in the Morphogenesis of Language Capacity

In higher apes (e.g., chimpanzees), one finds two means of social communication: grooming (lousing, caring for one another) and social calls. The (manual) grooming mode dominates, consuming about 20% of the budget. The critical transition (from the common ancestors of chimpanzees and men to australopithecines) was probably a dominance shift due to larger groups (cf. Dunbar [21]) and richer social connectivity in groups. Moreover, the loss of fur, probably due to an adaptation to long-range running during daytime under the sun, reduced the grooming functionality. This development already began 4 million years before the present (BP). As a result, phonic communication was more time-economic, and phonic contact with socio-emotional content largely replaced bodily contact.

Figure 6.1 illustrates this shift of dominance, which had to cross a point of symmetry.

Studies of the behavior of apes in the wilderness have shown that some species have a simple system of calls with referential functions. They allow the other members of the group to distinguish different dangers. These may stem from animals like eagles (attacking from above), carnivores (e.g., lions attacking below), and snakes (creeping

([32]: 22–26), a language faculty in the broad sense (FLB) is separated from a language faculty in the narrow sense (FLN). The latter is a computational faculty, including recursion and discrete infinity. As in the basic publication of Chomsky [14], language is reduced to a set-theoretical automaton, excluding all questions of reference and social function, the FLN is a "ghost construction".

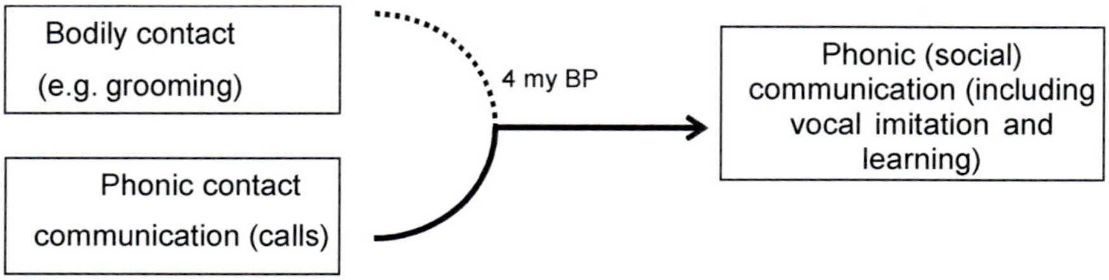

Fig. 6.1 The dominance shift made social phonic communication the central technique of social control and management

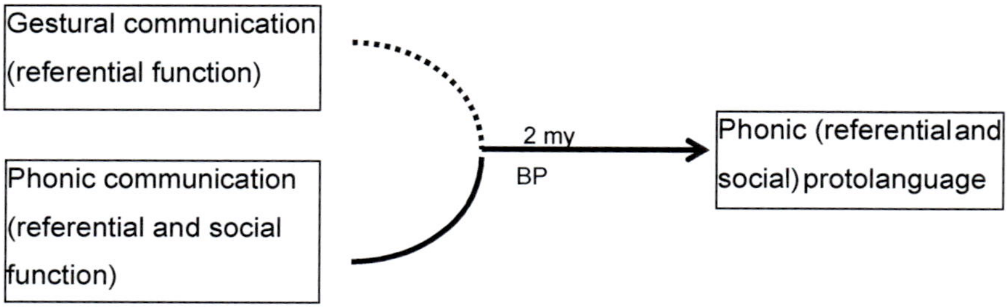

Fig. 6.2 The dominance shift from gestural to phonic in referential communication

in the trees).[8] Parallel to this simple system, bodily motions, gestures, and gaze directions give communicative cues, which allow for a spatial interpretation. They can thus be elaborated into a "language" of body postures and gestures. Therefore, the rich system of gestural signals was functionally parallel to a poorer system of calls.[9] For example, let us take the gestures of the hand. It is clear that as soon as hands are fully occupied with other functions like carrying tools and objects, or if communication occurs in the dark or at a distance (with obstacles between), the gestural "language" is ineffective. As such circumstances probably prevailed in the ecology of the australopithecines that lived in the savannah, and as the ear had to increase its capacity for discrimination due to the permanent danger of carnivores in the environment, the bimodality between reference by gesture and reference by phonic articulation shifted toward the latter. Figure 6.2 illustrates this transition.

---

[8] Hauser et al. ([32]: 31) argue that the calls lack intentionality and that the animals (velvet and rhesus monkeys) are only able to extract information from the acoustic signals. As we cannot interview monkeys, the question of intentionality must remain open. It would, however, be sufficient if a cooperative practice was genetically prepared and quickly developed in these species. Brain scanning results show that learning does not presuppose consciousness, thus even very low levels of consciousness in hominids would not preclude social learning of signs and their meaning, cf. Henke et al. [33].

[9] Kendon [38] shows that "some of the important capacities for the oral-vocal actions required for speech can be found quite widely in various non-human primate species, suggesting a long evolutionary history." (ibidem 73).

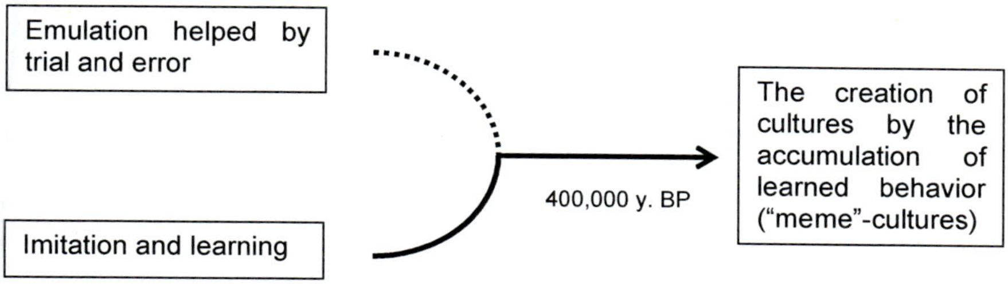

Fig. 6.3 The bifurcation which separates simple cultures based on emulation and "meme"-cultures[10]

The result of this functional evolution lies midway between biological and cultural evolution insofar as the repertoire of manual behavior in grooming and gestures and the repertoire of social and referential calls is acquired. Relatively to higher apes, the resulting protolanguage probably contained a rather large "lexicon" of social and referential calls (e.g., about 30 to 50 patterns), with different types of evaluative modulations (social calls) and categorical distinctions (referential calls). This stage prepares a referentially motivated sign architecture, i.e., language.

The capacity for imitation of phonic material, quick and stable memory entries and corresponding search procedures, and semantic network-building faculties presuppose a better-organized brain (cortex, centers of auditory and visual detection) and enough space (synaptic connectivity) to build a memory that associates phonic patterns with other (visual, olfactory) cues. This evolution leads us to a third bifurcation. It concerns imitation and learning in the case of motor behavior and symbol use. The baseline is defined by the presence of mirror neurons in higher primates and their capability of quick motor learning and motor control (cf. Rizzolatti and Arbib [52]) and the rise of a theory of mind in late hominins (e.g., chimpanzees). The bifurcation occurs between simple emulation and stable cultures with rich traditions. Simple "cultures" of tool use have even been documented for chimpanzees (cf. Boesch [10]). They are still linked to immediate success (reward) but prepare a more general strategy of imitation and learning from others without immediate pragmatic support (or "grounding"; cf. Cangelosi et al. [13]). In computer simulations, this distinction is called the "toil" versus "theft" strategy. In the case of symbolic learning, a label is either learned concerning its referent via trial and error or "stolen" from the symbolic behavior of others (the semantics are filled in later). Human infants systematically are "symbolic thieves" in the sense of these experiments. Human cultures accumulate information transmitted without being applied and tested extensively by every user. Dawkins called this information "memes"; cf. Blackmore [8] (Fig. 6.3).

The phonic protolanguage that integrated social and referential communication and was able to receive and transmit the accumulated cultural knowledge must have

---

[10] These bifurcation schemata were the content of a conference in Alicante and a publication in Spain; cf. Wildgen [80].

reached a first plateau, which was sufficient for the survival of this new species[11] and allowed its migration into Eurasia and its diffusion into Africa (Homo sapiens). The biological morphogenesis underlying the emergence of the language capacity of all human populations asks for this process's unusual rapidity and effectiveness. In the following section, we consider the hypercycle hypothesis introduced by Eigen and Schuster [22] in the context of the evolution of life on earth to explain the rapid evolution of human language.[12]

6.1.3 Autocatalytic Dynamics and the Evolutionary Hypercycle

The autocatalytic features are evident in the case of viral replication and variation, given the extreme efficiency of these systems. The simple catalytic cycle uses some catalyst (mediator) in the environment, and the rate of replication depends on the concentration of the catalyst. In the case of hypercycles, the system produces the medium (catalyst). As a consequence, the rate of replication does not grow exponentially. Instead, it grows hyperbolically, which allows for rapid evolutionary processes. However, hypercycles have other essential features:

- If different replication systems co-occur, the hypercycle can bind them together, eliminate rivaling (not bound) processes and stabilize the cooperative system (the hypercycle is a synergetic system; cf. Haken [29]: 315 f.).
- The whole cycle acts like one unit, rivals other hypercycles (if they exist), and eliminates less organized processes.
- The single systems bound by the hypercycle can still work with low complexity, but the cooperative system has access to higher levels of complexity.
- The hypercycle can reach optimality without external selection pressures.

Despite these advantages, hypercycles are endangered by changes in the external milieu. Suppose the biological or symbolic system survives for some period. In that case, it can trigger an all-or-none decision in the evolutionary line because all alternatives have been eliminated, and further evolution cannot return to an earlier situation. The evolution must follow the direction chosen during the operation of the hypercycle. In a sense, the hypercycle reduces the potential for selection and determines long-term evolution. It leads to a sequence of steps that seem to follow

[11] Although we know nothing about communication in Homo erectus, the principle of evolutionary continuity (on which Darwin founded his theses) motivates the hypothesis of a *protolanguage*, i.e. a way of phonic communication that prefigures the language typical for our species. For example, it could correspond to children's one- and two-word utterances in their early second year of language acquisition or rudimentary pidgins.

[12] This hypothesis was first presented at the conference "The Cradle of Language" (Stellenbosch, South Africa, 7th to10th of November 2006) and published in an elaborated version in Wildgen [77].

logically from an underlying principle. It looks like an intelligent design, although it only exploits natural processes and obeys the laws of nature.

Non-coupled self-replicative units guarantee the conservation of a limited amount of information that can be passed on from generation to generation. This proves to be one of the necessary prerequisites of Darwinian behavior, i.e., selection and evolution. Similarly, catalytic hypercycles are also selective, but in addition, they have integrating properties, which allow for cooperation between otherwise competitive units. Yet, they compete even more violently than Darwinian species with any replicative entity not being part of their own. Furthermore, they can establish global forms of organization as a consequence of their once-for-ever-selection behavior, which does not permit a coexistence with other hypercyclic systems, unless these are stabilized by higher-order linkages' (Eigen and Schuster [22]: 6).

The most dramatic difference between physiological microevolution and symbolic (linguistic) macroevolution concerns the fact that the first one stores and activates the phylogenetic memory of a species, whereas the latter stores the historical/cultural and the individual/biographical memory. The exciting feature of catalytic and hypercyclic organizations is that they enable faithful replication and dramatic selection by their hyperbolic growth. This means that all types of organizations, not part of an operative hypercycle (i.e., all competitors at a lower level), are repressed. In the following, this promising but abstract model is filled with further details from human evolution. One can choose two application routes: cognitive (neural) and social (cultural). The cognitive and the social route enter a cycle of coordination, which tends to induce individuals to select cultural content as cognitive content and to eliminate much potential content which is not socially relevant. This strongly selective (hyper) cycle may be called socio-cognitive. In the two periods in which new behavior surfaced (cf. at the stage of Homo erectus and Homo sapiens), the socio-cognitive hypercycle has selected humans for symbolic competence. In the co-evolutionary system between a cognizable ecology and symbolic competence, the following hierarchy is plausible:

(a) Already in the last common ancestor of chimpanzees and humans (LCA), contextual space acts as an external memory of affordances, which is indexically given by paths of social locomotion and predator/prey-locomotion, harvesting locations (and times), dangerous locations, places for sleep, courtship, housing, and frontiers of territories. Moreover, these indexically loaded areas and places function as a catalyst of social action as they coordinate social perception and action.
(b) As soon as social space is more organized explicitly concerning its perception and social use, it unfolds in a cycle of social "investment". Architecture and the spatial organization of a village (or later a town) are clear examples. This level is autocatalytic as the spatial organization becomes a cyclic structure in which different functions cooperate. Figure 6.4 sketches such an autocatalytic cycle.

In each subspace, specific symbolic media are evolved. Thus, the painted Paleolithic cave (in the Franco-Cantabric culture) is a specification of the mythical/ritual space and is also connected by its illusionist paintings to the outside space of hunting. The relation is iconic, indexical (in its magical impact), and symbolic (in its abstract

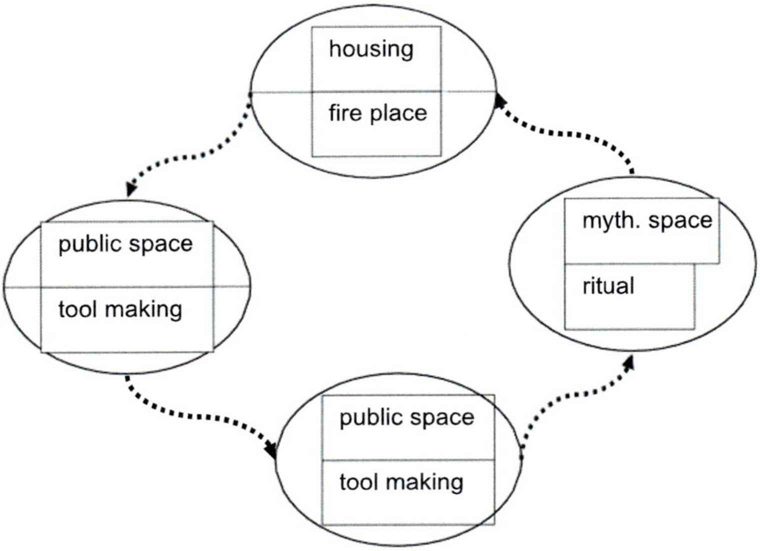

Fig. 6.4 Symbolically invested subspaces (above) and possible symbolic functions (below)

signs; cf. Wildgen [74]: 80–83; [75]). The dark, painted cave points to cave openings, and later, huts where people live. The open space in front of the cave openings or huts is a public space where the production of artifacts and the distribution of shared food occur. This public space borders the open field of chasing and harvesting. Human action patterns occur inside a specific space or make the transition from one space to the neighboring one. In rituals, the core of these action patterns is fixed. The coding of action patterns by rituals is a social preparation/presupposition for linguistic rules/grammar.

In further development, a new level of symbolic consciousness is reached when different symbolic modes (e.g., languages and myths) clash, e.g., in the large Neolithic societies of Egypt and Mesopotamia. The single fields in Fig. 6.4 reorganize in a hypercycle that produces a new, standardized symbolic system, e.g., a codified religion and a written language. Possibly the Franco-Cantabric culture (35,000–13,000 y. BP) and the Sahara cultures (later) had already reached this level. However, as the code of their abstract signs cannot be deciphered, this hypothesis cannot be substantiated.

In the course of cultural evolution, the effects of integration imply a network of symbolic forms. Symbolic forms are the manifestations of social knowledge, and language is the most prominent symbolic form that codes social knowledge. At the "higher" levels of learning, processes of self-organized reconstruction play a decisive role. Specific institutions had to be created to stabilize the social knowledge level or even increase it. The new codes called "religious code" and "written language" are at the heart of such institutions (cf. Chap. 5).

The symbolic forms are multiple and, in their specific elaboration, they are not species-universal, i.e., every separated community develops different symbolic systems, for instance, languages and dialects. Nevertheless, they are comparable via common principles of morphogenesis applied at different ontological levels (cf. Chap. 7).

6.2 The Semantics of Space and Time in a Protolanguage

One can distinguish two aspects: processes in space, such as spatial orientation and navigation, and temporal classifications and rhythmical patterns. The representation of *space* has to do with frontiers (their transition) and perspectives. A first perspective is centrifugal, i.e., starting from the self and its bodily motions and locomotions, an 'experienced' three-dimensional space is cognized: in front of–behind (go), above–below (climb, fall), and left–right (grasp with the left hand or the right hand). This space of bodily motion with feet and arms defines the immediate space where objects may be approached, reached, and manipulated. The intermediate space depends on man's ecology; it can be the housing; first, the cave, the shelter ("abri"); later, the village; the distal space contains roughly all possible itineraries (of hunting/gathering). The second perspective is centripetal, i.e., the self is seen as the place of effects triggered by external causes. The sky, the horizon (specific points where the sun sets or rises), the favored direction of winds, and the ridge of mountains may be the external locus of orientation for the self, who is at the center of a force field implicit in these delimitations. Many myths and religions refer to this extreme locus of orientation as they interpret the fate of humans as standing under the control of such distant (and often invisible) forces. The cognizing of such schemata for orientation may only show up in behavior (as it does in many animals), it may be gestured, or it can be deictically organized in a phonic language.[13] For the Homo erectus, the cognizing space is clear. The inner space is defined by hands and instruments, and the medium space by choice of dwelling places (to which the group could return). The centripetal organization is involved in long-range excursions and migration. As the orientation system cannot be genetically coded, it must be learned, adapted to changing contexts, and socially shared. Language is one possible solution to this problem, be it gestural (behavioral) or phonic. As humans have chosen the path of phonation, it is plausible that our ancestors began to proceed in this direction.

A protolanguage must categorize events and actions (by proto-verbs) and discriminate stable entities (by proto-nouns). The question arises as to whether temporal, dynamic, quantitative, and qualitative *relations* between them can be mastered, and if so, to what degree.

[13] Cf. the research on types of orientation in different ethnic groups, e.g., research conducted at the Max Planck Institute of Psycholinguistics, Nijmegen by the group of Prof. Levinson; cf. Levinson [43].

The manufacturing of stone tools (and *a fortiori* of tools shaped with the help of stone tools) goes cognitively beyond the basic grasp scenario. One hand (or one foot) must fix the pebble, and the other hand grasps the stone or bone which hits the stone. Finally, the planned breaking off subtracts material from the chosen stone and produces the desired sharp edge of the pebble after several strokes. This scenario involves two objects, two hands, and a change in the shape of the pebble (the separation of parts from it). Additionally, it manifests a branching sequence and iteration characteristic of the syntactic organization in natural languages (cf. phrase structures and repeated embedding).

6.3 The Morphogenetic Transition Between a Protolanguage and Full-Fletched (Modern) Languages

Our empirical knowledge concerning languages is based on the observation of historical languages documented in writing (e.g., Sanskrit, classical Greek, Latin, or Arabic) and the research on living languages, dialects, and sociolects. Therefore, hypotheses concerning unwritten, prehistorical languages and a fortiori, the stadium of protolanguage, must extrapolate our knowledge about known languages and refer to more general principles known from biology or neurology. Therefore, such hypotheses are just well-informed guesses. Nevertheless, they contribute to a global view of human cognition and culture and are relevant.

6.3.1 The Self-Organization of a Grammatical System

Self-organization is a principle formulated in the framework of cybernetics (Ashby [5]) and involves the search for a stable state in a deterministic system. As already programmatically expressed by Norbert Wiener [65], it is extrapolated from physical to biological, eventually symbolic systems. Moreover, in morphogenesis, equilibria and attractors (stable states) are also central notions. Therefore, self-organization and morphogenesis follow a similar strategy of explanation.

Kirby [39] argues that compositionality (and thus syntax) can already emerge if the size of the lexicon (meanings associated with linguistic expression) increases. “The number of meanings covered increases dramatically, as does the size of the grammar” (ibid. 313f). Steels [56] simulates language evolution based on evolutionary games. A stable state emerges when the number of meanings increases, and due to organizational economy, the size of the grammar drops. Given the lexicon’s specific size and the transmission dynamics (learning), languages tend to evolve a rather general syntax without any pressure from environmental or sexual selection.

Fitch ([25]: 385) is skeptical about the relevance of computer simulation for studying evolution. He writes: "In practice, however, the demonstration of a theoretical possibility does not by itself, tell us how the pattern was 'discovered' evolutionarily". This is true,but if we cannot observe or find documents of such a discovery, we must still choose between different theoretical possibilities. The simulation provides better arguments than aprioristic deductions from definitions like "recursive power" or the necessary existence of universal grammar (in Chomsky's tradition).

The purely syntactic problem of chaining elements of an existent vocabulary does not require a specific endowment and evolutionary processes enabling it. The real problem is semantic compositionality because the composition or blending of spaces with different topologies and the account of the dynamics inherent in verbs is crucial for sentential units. This tremendous problem must be resolved to allow stable and reliable communication via phrases and sentences. To arrive at a conventionalized system of syntactic behavior, early humans had to consider two major factors:

- The cognitive demands for a stable solution of semantic compositionality,
- and the communicative and social demands for a compositional level of referentiality.

The solution to this problem is the *gain* of the evolutionary game called human language.

Even if the cognitive capacity was given, human society must still have a strong demand for high performance. Probably rewarding situations often arose by chance, and the evolving species spontaneously used the "dormant " capacity. As soon as a protolanguage was developed, it brought about long traditions of language usage up until modern times. The central question is not how syntax came about but what made it rewarding to use the available cognitive potential for syntax. The payoff can be a social or an individual one (which can lead to higher social competence and thus to social gain). A plausible model for such higher communicative demands due to social evolution is still missing.

6.3.2 Further Steps of Complexification in Language

Sentential patterns may be elementary even in human languages, e.g., in pidgins, in learner languages, and even in standard languages with broad usage (e.g., the so-called minimal languages analyzed by Gil [27]). Human utterances are, however, not restricted to isolated sentences. On the contrary, natural units are sequences of sentences, so-called turns in conversation, adjacent pairs as in question–answer, and narratives or arguments. A fundamental problem concerns the stratum of language (from phonology to discourse) to which a selection process applies. As this is usually the level of holistic behavior, we presume that textual behavior is the proper level on which selection effects play a role. Therefore, human evolution must have been selected for the effective use of language in social communication and not at the level of sentences or words. These levels are only selection relevant insofar as they allow the construction of coherence in narratives, descriptions, or arguments. The increase of the lexicon and the availability of case-frames (action–schemata) and spatial categorizations establish the source domain, in which a very complex grammar system could emerge by self-organization. Another source domain in the morphogenesis of phrasal and sentential syntax is the capacity for producing and understanding a rapid sequence of phonic events due to short- and long-term phonetic memory. Studdert-Kennedy ([58]: 17) says: "Without a pre-adapted system for storing phonetic structure independently from its meaning, syntax could not have begun to evolve." Man's essential syntagmatic (sequence controlling) capacities are evident in morphological and phrase compositionality. The complexity of syntax seems overwhelming if we consider modern written languages that are the focus of most linguists. However, natural languages can be elementary on the syntactic level. Therefore, it is not necessary to consider this complexity as a general characteristic of human languages. They can achieve such complexity, but this is not necessary. Comrie and Kuteva ([17]: 202) conclude from their comparative analysis that "human language(s) might have been much simpler and highly functional, and might have lacked grammatical forms such as case inflections, agreement, voice markers, etc."

6.3.3 Summary of the Evolutionary Morphogenesis of Human Languages

The *biological* evolution of human language is a continuous process in which bodily preconditions are decisive. In the further *social* evolution that started with the increase of group size (and group organization) and new forms of symbolically ruled social behavior, bifurcations with symmetry-breaking and convergence occurred: from manual to phonic contact management, from gestural to phonic deixis and referential location, and from context-dependent emulation to cultural learning. After these dramatic shifts toward a complex phonic communication system with socio-emotional *and* referential functions and cultures based on symbolic transmission, the centrality of language for human survival and expansion was firmly established. Language capacity became a species-defining character of humans.

The advanced stone-age industries show that Homo sapiens crossed this barrier before they began to move beyond their limits in South-Eastern Africa (200–100,000 y. BP) and to migrate out of Africa (between 100,000 and 70,000 y BP); cf. Wildgen ([74]: Chaps. 4 and 5) and Wildgen [81].

The significant effects on language were:

- A more extensive and steadily growing lexicon;
- the mastery of rapid and complex strings of phonic signals and corresponding functional-semantic patterns as shown in lexical innovation, composition, grammaticalization, and complex phrasal syntax;
- a new level of creativity in language and art linked to the growing complexity and diversity of languages.

6.4 Morphogenetic Schematization in the Lexicon of Natural Languages

The following sections give a summary of a morphogenetic model of human language. The focus is not only on specific biological processes in the actual formation of a living being (e.g., gastrulation, cell division, and the genetic or epigenetic control of maturation and growth), but also on the existence and the further unfolding of abstract "morphogenetic" forms in the sense of Kant's schema-theory, Goethe's "Urformen", Saint-Hilaire's "structural plan".

Modern evolutionary biology accepts the persistence of a set of ancient genetic factors (the "homeobox") responsible for a kind of "unity of composition" observed by Saint-Hilaire in comparative anatomy (see the rise of "Evo-devo-models"). We cannot discuss these issues in evolutionary theory (cf. our remarks in Sect. 1.1.2). The underlying philosophical and theoretical position was already formulated in René Thom's book (1972, translation 1975: Structural Stability and Morphogenesis), and its consequences for linguistics and semiotics have been specified in Petitot [48] and Wildgen [66, 68, 70, 72].

6.4.1 *Morphogenetic Principles Versus Universal Grammar*

The topic of universal grammar or inborn ideas surfacing in an incomplete and deformed way in human languages goes back to antiquity and is lined to the topic of the naturalness of language contrasted with the appearance of arbitrariness (conventionality); cf. Platon's dialogue *Cratylus*. Descartes' epistemology and the theory of grammar and rhetoric of Port Royal assumed a logical blueprint of human language (logic understood as a theory of human thought). Chomsky assumed this historical position to be the backbone of his theory of Universal Grammar, based on set theory and (free) algebra (cf. Chomsky [14, 15]). In the seventeenth century, Leibniz had already criticized the rationalistic position of Descartes and advocated a continuous and dynamic view of universals and grammar. Leibniz's ideas for universal geometrical characteristics of human thinking and creativity led to the modern topology. Poincaré's philosophy of science and the applications of results in differential topology by René Thom and Sir Christopher Zeeman (under the label "catastrophe theory" in the seventies and eighties of the twentieth century) established a theoretical framework for dynamic semiotics and linguistics, cf. Wildgen ([69]: 31f.) for a short historiographical discussion. We take as our starting point the natural sciences and mathematics before and after the millennium (2000), specifically the morphogenetic theorizing by René Thom.

Starting from morphogenesis in the domain of nature (for instance, geological and biological processes), René Thom has proposed formal schemata derived from catastrophe theory that generalize major types of morphogenetic evolutions. In the grammar of human languages, such process types show up as invariant form-giving principles that constitute a "homeobox" constitutive for homologies between human languages. We do not assume the existence of universal grammar (based on innate "ideas") but a gradient in the invention and selection of grammatical forms that leads to the statistical prominence of certain types of grammatical organization (constructions). This does not mean that grammatical structures whose genesis can take thousands of years are the immediate outcome of specific morphogenetic processes. One must instead assume myriads of communicative acts involving entire populations that create ad hoc lexical ad syntactic patterns, select among a given inventory, and change them in detail. The cumulative outcome of these processes selects and optimizes certain organization features in the lexicon and the syntax.[14] The details of such a long-term process with many degrees of freedom cannot be reconstructed in linguistics (in the present state). Therefore, the morphogenetic analysis of grammar must take a detour and use the morphogenetic patterns abstracted from nature and seek plausible patterns discovered in languages until now that are possible realizations of these abstract morphogenetic structures. For a detailed analysis of catastrophe theoretic semantics motivated by René Thom's proposals, see the analyses in Wildgen ([66, 72]; in English; [68, 70] in German; [73, 82], in French) and Petitot [48–50].

Finding the most basic entities underlying a set of complex structures is similar to the search for basic figures in geometry. The analogy between geometry and psychobiological phenomena was already known to Aristotle (in his treatise on psychology, "De Anima", Aristoteles, [4]: § 414b16). The abstract geometrical character of elementary representations is also evident in neurophysiology (see Orban [46]) and in the psychology of vision Kosslyn [40], cf. for the discussion of this research and its relation to Thomian thought Petitot [51]. The analysis of sensory inputs consists of mappings from a three-dimensional input into a precise control of activity in space and time. The mapping must conserve essential topological and dynamic characteristics and can forget metrical details and variations of a type of object or event. Therefore, the problem of *structurally stable mapping* lies at the heart of every theory of representation and semantics. The crucial result in this field is the theorem by Whitney that says: We can only find three types of points (all other types become identical to these if perturbed):

(1) regular points (Morse points); they do not qualitatively change under perturbation; we may say that they have a static identity (of self-regulation),
(2) fold-points (a frontier line between a stable and an unstable domain appears),
(3) cusp-points (two stable attractors conflict and one may appear or disappear).

[14] Evidence from language change documented by historical and comparative linguistics indicates that stability is maintained against forces of destabilization or new optima are sought. For example, Steels [57] simulated the language change from Old High German (ca.800–1000 AD) to New High German (ca. 1650–today) using computer agents in a language game. He concludes, "Although there have been phonetic processes (unrelated to function) that have eroded endings and merged forms, only those solutions that lead to a more optimal system from the viewpoint of semantics, morphosyntax, and phonology have undergone positive selection."(ibidem 348).

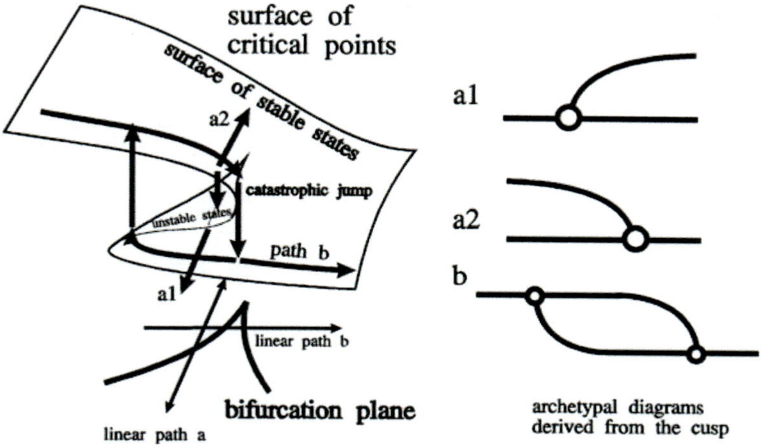

Fig. 6.5 The derivation of archetypal diagrams from the "cusp"

The classification theorem of catastrophe theory expanded this list to the cuspoïds and umbilics. After 1978, the embedding of umbilics in the double cusp was added (cf. Wildgen [66]: 81–92).

A further notion must be informally introduced: the linear path in an elementary unfolding.[15] If we consider linear paths in an unfolding, we can classify types of process schemata called EMISSION, CAPTURE, and (bimodal) CHANGE. They are derived from the catastrophe set (set of extrema) of the cusp. The diagrammatic simplification at the right of Fig. 6.5 eliminates the lines of (unstable) maxima; the circles symbolize the bifurcation points.

Thom proposed considering only catastrophes with a (co-)dimension equal to or lower than the dimensionality of spacetime. The basic scenarios of change and process in the cusp have two attractors. The butterfly has three attractors, and the elliptic umbilic has four attractors. Examples of these formal process types are given in the following sections.

We shall concentrate on the levels of lexicon and sentential syntax. Applications of the morphogenetic paradigm to phonology have been published in Wildgen ([70] in German and [72] in English). Applications in text linguistics, narrative analysis, and discourse are not considered here due to lack of space (cf. Wildgen [67], Wildgen [72, 73]: this chapter; in French).

[15] In the simplest case, the unfolding of a dynamical system under deformation (noise or perturbation) has a gradient dynamic without oscillations or chaos. This assumption allows the classification of all the structurally stable evolutions of the system. The practical consequences of the classification theorem are called "catastrophe theory".

6.4.2 Morphogenetic Structures in the Lexicon of Verbs

The lexical category of verbs that logicians traditionally neglected[16] becomes the theoretical core of a morphogenetic analysis. The first questions that the categorization of a process, an event, or an action raises are:

- What invariant structures underly the processes in question such that a lexicon of verbs can stably refer to them?
- What are the motor controls and perceptual patterns that mentally appropriate such processes?
- Finally, what mental representations link the perceptual-motor correlate and the linguistic forms?

We start from the psychophysical interface, i.e., the perceptual and motor systems, which establish a link between the world and the human body. As the work of Gibson [26] and, more recently, that of Haken [30, Kelso 37], and others show, the qualitative laws of external physics control this interface. The motor programs modify the autonomous dynamics of the body's extremities and their contact with objects (e.g., the floor for the feet). These autonomous dynamic structures determine variables for the perception of movements and the cognition that regulates these systems. It can be concluded that the brain reflects the external dynamics (by adding other parameters to it and distorting it in its metric). However, the question remains: Does this coordination with external physics also control the higher levels of cognition, especially linguistic cognition? We assume an intermediate level, called imaginal or schematic. It applies processes that become increasingly independent from the psychophysical grounding and more context-dependent (as a consequence, they depend on chance). In what follows, we will start from the psychophysical level to find schematizations (imaginal representations) that underlie the semantics of verbs. We distinguish three typical levels of organization in the lexicon of verbs and verb phrases (the lexicon of nouns and noun phrases will be the topic of the next section):

a. Locomotion and its linguistic schematization,
b. the control of an object by an agent,
c. and the interaction between agents.

ad a: The morphogenesis of the cognitive and semantic schematization of motor acts.

The psychophysical perspective has the advantage of creating a link between the cognitive, the sensory-motor, and the dynamics of the external world. Movements have two levels of control):

a. Control of the topology of the movement and the coordination of the different moving parts.
b. Control of the metric of the movement. It gives the necessary precision for a concrete situation/environment.

---

[16] See for example the logic of Port Royal which wanted to reduce this part of the lexicon to the single verb "to be".

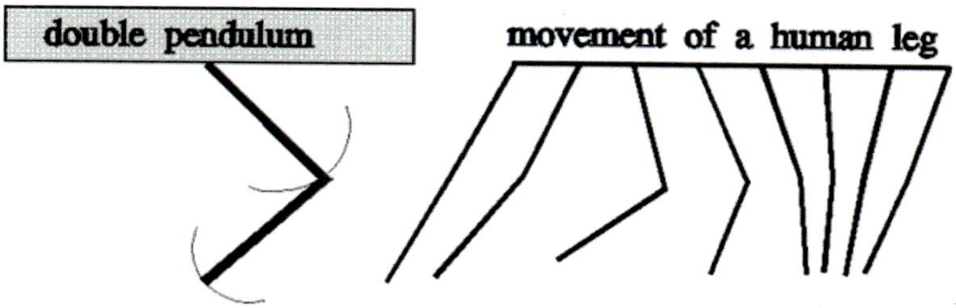

Fig. 6.6 The movements of the double pendulum and of a leg in walking

The first level eliminates factors that blur the general approach of a target. In this sense, the coarse (topological) control is locally teleological. Metrical control, on the contrary, has the effect of adjusting the movement and eliminating insecurities and vagueness.

In the case of the movement of the body's extremities (e.g., legs, arms), one can use the physical description of the pendulum (double pendulum) as a fundamental schema. Figure 6.6 shows the correspondence between the double pendulum and the walking motion. The movement of the body supported by the hips is in coordination with the relative movements of the thigh (measured at the knee) and the leg (measured at the ankle).

By moving from the rest position, the leg moves toward the goal, the new attractor. The iteration of local movements allows for a quasi-continuous movement. The rhythm of it can further specify the movement or even identify the agent of the movement. The movement zones with a very high degree of coordinated control are, at the same time, the domains that organize semiotic expression, for instance, the facial muscles and the movement of the hands and fingers. The complexity of motor programs is not directly related to lexical distinctions because the latter only classify recurrent types of movement in a much coarser way.

Specific movements are directed toward an attractor (a target). This orientation implies a separation of the starting and the ending point. This bimodality corresponds to the fundamental transition in space and its lexical correlates. Example: enter/*exit* or *come/leave*. Two essential types of elaboration must be considered:

(1) Instrumental elaboration. The control of the body's limbs and the coordination of complex movements can be modified or specified by inventing and using instruments (and machines).
(2) Causal elaboration. Further technical control uses the physical causalities discovered and implemented in science and technology.

ad b: The morphogenesis of an agent's control over an object.

In the intentional act of an agent directed at a less intentional object or entity, two aspects can be distinguished:

(a) The configurational aspect describes the topological and kinematic relations between the agent and object.

(b) The energetic (or intentional) aspect introduces the force of the agent and the effect of this force on the object. This force is first psychic; secondly, it has physical effects.

In their classification of the lexicon of German verbs, Ballmer and Brennenstuhl [6] distinguish, at this level of complexity, two groups of verbs:

(1) The creation, destruction, and regeneration of objects (elements of the environment).
(2) The subject's impact on the state of objects and subjects in its environment.

For group (1), it is easy to see the correspondence with the emission and capture schemata in catastrophe theoretical semantics (cf. Fig. 6.5 in Sect. 6.4.1). In the lexicon, the corresponding verbs are, in most cases, divalent, as in:

EMISSION
Albert tells a story
Berthe calculates the result
CAPTURE
Charles eats the soup.

Often the semantic type of the produced objects is incorporated into the verb, as shown by the following German verbs (EMISSION type):

- schneidern (to sew)
- töpfern (to make pottery)
- texten (to write texts).

The subgroup called regeneration/disaggregation by Ballmer and Brennenstuhl [6] refers to a space of qualities. We will look at some examples:

(a) verbiegen (distort, twist)
(b) reinigen (cleanse).

The processes that are classified by these verbs refer to a qualitative space with the following states:

1. right > twisted German: verbiegen (deform),
2. clean > dirty German: reinigen (English: to clean; French: nettoyer).

In the control space of the cusp (cf. the folded surface in Fig. 6.7), we have paths that go from:

(a) $(+) \rightarrow (-)$: verbiegen (to distort)
(b) $(-) \rightarrow (+)$: gerade biegen (to straighten).

Figure 6.7 shows the dynamic modeling of the French verbs: nettoyer (clean) and salir (make dirty) and corresponding adjective scales: propre (clean) and sale (dirty); cf. Wildgen ([73]: 99 and Chap. 3).

ad c: The morphogenesis of the cognitive and semantic schematization of interaction.

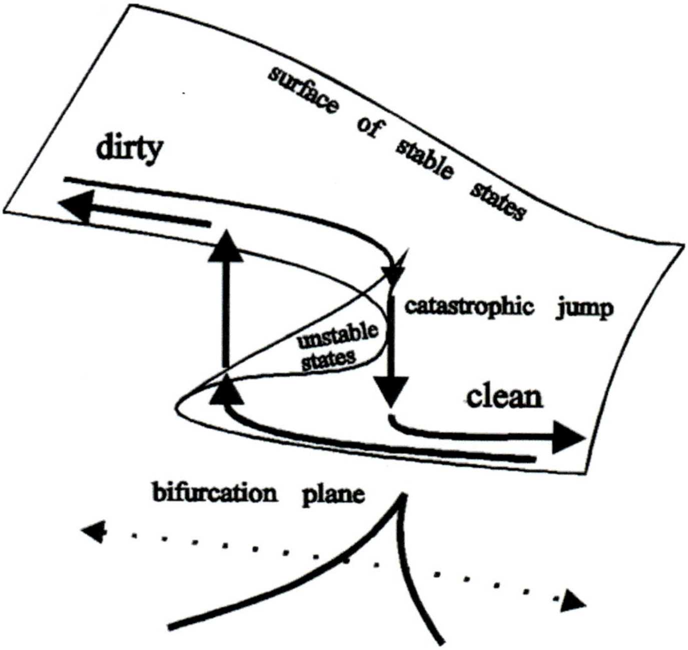

Fig. 6.7 The dynamics of the English verbs clean versus make dirty and the corresponding adjectives clean and dirty

An interaction scene that connects several human agents already presupposes a very complex perceptual and conceptual analysis in the individual observing the scene. Above all, it presupposes a degree of social perception that goes beyond the simple control of the action by the agent himself. For example, we know that primates can consider the perspective of another ("decentration") to a degree comparable to the decentration manifested by a two-year-old child. From a repertoire of action controls, one can reconstruct possible patterns of social interaction. However, it turns out that only a small group of these coordinated interactions achieve high stability, allowing schematization and semantic classification. This restriction requires an explanation.

From the angle of the spatial configuration, we can describe the gift, i.e., the scene during which two people exchange an object, by the topology of the attentional focuses. Petitot ([50]: 272ff) elaborated on a proposal by Christopher Zeeman for a model that uses cognitive algorithms, such as cut locus and diffusion contours. In this elaboration toward neuro-vision, the catastrophe theoretical model gains more theoretical depth because it shows that the mathematics of differential topology can be used in the specific context of visual scene analysis. The semantic archetype would, in this perspective, be first the result of morphogenesis in visual pattern recognition. As such, non-human primates could have possessed this faculty. Then, this pattern would have gained social and cultural significance in the ritualization of gestures.

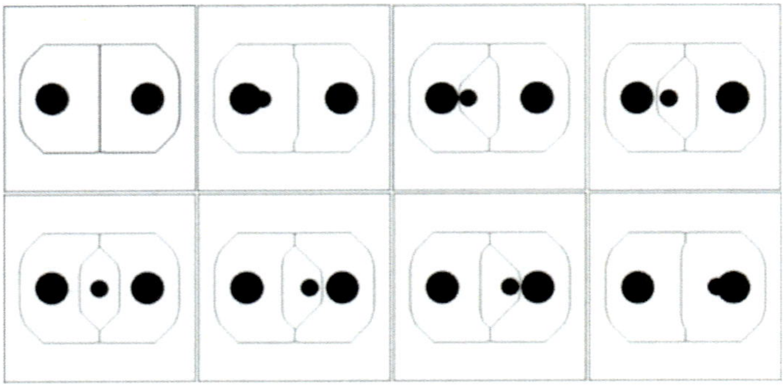

Fig. 6.8 Phases of the schema of transfer (give; from the left upper corner to the right lower corner) applying principles of neural dynamics in vision ("cut locus" analysis) in Petitot [50]: 274: Fig. 4.8)

Finally, with the transition to phonic language, evolutionary continuity would have created a stable reference pattern in early human communication (Fig. 6.8).

Petitot ([50]: 273) writes: "the temporal evolution of the cut locus itself is slow dynamics [...] and may present bifurcations, emergence, and vanishing of branches, or splitting of branches. These dynamics encode events of interaction between actants. We can, in this way, develop a program analogous to contour diffusion. Figure 4.8 gives an example of the transfer type".

The middle phase (lower left corner) coordinates the activities centered on two individuals in the starting and the goal positions; it is also the most unstable point of the whole process.

This scheme of interaction is transformed into language schematization by transfer verbs. The energy asymmetry defines an intentional direction. We can distinguish:

- the source agent that initiates a process;
- the object that changes the possessor going through a change of control and dominance;
- the target agent, the one who holds control of the object at the end. This state is the goal of the intentional action of M1, and it establishes, at the same time, an asymmetry, which may initiate M3 to fulfill a reciprocal exchange (Fig. 6.9).

The second large field of lexical entities has a nominal character and is classically labeled as nouns, adjectives, appositions, pronouns, and relative clauses.

6.4.3 Morphogenesis and Attractor Dynamics in the Lexicon of Nouns, Adjectives, and Other Nominal Attributes

The lexicon of human languages has a high degree of arbitrarity. It was evident to the comparatists of the nineteenth century, and de Saussure called this principle

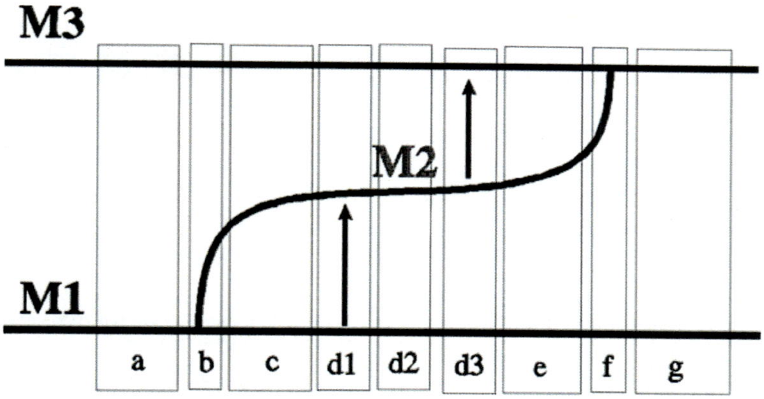

Fig. 6.9 Catastrophe theoretical schema of the transfer archetype "GIVE" with the critical phases d1, d2, and d3

"l'arbitraire du signe" (the arbitrariness of the (linguistic) sign). Although morphological and syntactic patterns of the languages in a family of languages can be stable over long periods, the lexical entities diverge very quickly. Even the dialects of the same language differ dramatically in the labels for plants, animals, and artifacts. This variability became evident in the systematic compilation of geographic variation in the atlases of languages established at the end of the nineteenth and the beginning of the twentieth century.

In lexicology and lexical semantics, since the twenties of the last century, models of field linguistics came to the fore; the notion of "field" implies gradient dynamics. However, elementary patterns of lexical semantics were already known in antique (Aristotelean) logic and rhetoric. For example, Aristotle used in his taxonomy of plants and animals the technique of specific difference between a general term and a more specific one: "genus proximum et differentia specifica". This technique can be translated into a feature notation, i.e., a term higher in the hierarchy may be defined relative to the genus proximum, the next lower term, via the specific difference between them.[17] This idea was taken up by Katz and Fodor [36] and used in generative grammar after 1965. We shall exemplify the technique based on its usage in Labov [41], who elaborated on it in an empirical (sociolinguistic) context.

The meaning of bowl can be analyzed regarding a set of five features (defining predicates):

- feature 1: diameter; weight w_1,
- feature 2: height; weight w_2,
- feature 3: the existence of a handle; weight w_3,
- feature 4: use (food–non-food); weight w_4,

[17] Darwin appreciated Aristotle's contribution to biology (in 1882) and D'Arcy Thompson translated Aristotle's History of Animals in 1910. René Thom gave the under title "Physique aristotélienne" to his book on Semiophysics (1988).

- feature 5: material; weight w_5.

The features have different weights, and this weight may even depend on the contexts in which the word "bowl" is used. For example, as Labov [41] shows, the weight of feature 4 is higher in the context of "kitchen" and "restaurant", whereas in the context of craftwork and art, features 3 and 5 are more relevant. Fuzzy semantics (cf. Zadeh [83]) have generalized the use of "weights" as degrees of membership of an element in a set with values between 0 and 1 [0,1]. As the scale of values is continuous, we can define maxima and minima on the scale, i.e., use attractor dynamics.

Even the hierarchical branching of lexical items can show attractor dynamics. The ethno-taxonomies analyzed by Berlin [7] and Rosch [53] show the prominence of intermediate (Berlin) or basic (Rosch) categories. Thus, the hierarchy in the series *tool, hammer,* and *claw-hammer* has the term hammer as its center. It is the attractor of this array; cf. Sect. 2.1.3 in Wildgen ([76]: 40–42; in German).

In antique rhetoric, two other relations between lexical items were distinguished: metonymy and metaphor. This tradition was transmitted without significant losses to modernity and is the starting point of the semantics of metaphors in Lakoff and Johnson [42]. Thus, the sentences:

- argument is war (metaphor)
- the part stands for the whole (metonymy)

define a type of semantic transition that can be applied to many words and utterances which fit both arguments of the metaphorical/metonymical relation (cf. ibid. 4). Examples[18]:

The metaphor:*argument is war* may be articulated in sentences like:

- John's *claims* are *indefensible,*
- His *criticisms* were right on *target,*
- He *shot down* all my *arguments.*

The metonymy: *the part stands for the whole is* articulated in sentences like:

- We don't hire *longhairs (longhairs stand for someone having long hair),*
- The Giants need a *stronger arm* in the right field (a strong arm stands for a player with strong arms).

However, some of Lakoff's and Johnson's results were basic knowledge in gestalt psychology, especially in "attribution theory", since the fifties of the twentieth century. What is new in Lakoff and Johnson [42] is the role played by locutions and proverbs like *time is money* and *argument is war.*[19]

---

[18] The "semantics of metaphors" initiated by Lakoff and Johnson is critically analyzed in Wildgen ([76], Chap. 3, pp. 65–90; in German).

[19] The orientational function of fixed locutions or frequent images points to Jung's earlier theory of psychological archetypes. C.G. Jung collected symbols from many cultures in the form of pictures and sculptures and compared them to pictures produced as individual transcriptions of dreams and visionary experiences. From the comparison of these materials, he concluded that there is a level

The flaw of lexical semantics lies in the subjectivity (on the part of the analyst) of all analyses. Charles Osgood (1916–1991) developed and applied a tool to analyze lexical meanings in a near-to-objective technique. It consists of choosing a set of polar adjectives in a given language and having experimental subjects rate the relative position of lexical items on a seven grades scale. The advantage of this method is that the subjective character of meanings is the target of the analysis and does not interfere with the subjective evaluation or prejudices of the psychologist or linguist as an analyst.

The multivariate statistical analysis leads to a low-dimensional construct of factors underlying the correlations between the tested items. One of several optimal factorizations is given the typical loading of relevant scales on three factors called Evaluation (E), Potency (P). and Activity (A). Osgood et al. [47] showed that this three-dimensional semantic space has a biological meaning, i.e., the general factors reconstructed statistically refer to basic dimensions of human behavior that underlie the lexicon of adjectives and nouns. The morphogenesis of the rich semantic space of a lexicon can be considered as the unfolding of a low-dimensional, biologically-based semantic space.

In the transition between the lexicon and the syntax of languages, there exist processes that are labeled as "grammaticalization", i.e., items belonging to the lexicon are transferred into items having a syntactic function. The opposite direction is also relevant; for instance, relative clauses derived from sentential structures acquire the character of adjectival determinations or restrictions in a noun phrase. The morphogenesis of case markers is one example of grammaticalization. As it is relevant for our analysis of valence patterns in Sect. 6.6, it will be discussed in more detail (cf. Wildgen [82] for a fuller account in French).

6.5 Morphogenesis and Grammaticalization (Applied to Case Marking)

The world's languages show astonishing diversity in the morphology of words, i.e., in suffixes, prefixes, and infixes. The basic process, i.e., the transition from (free) lexical entities to bound morphemes, is called grammaticalization. This process means bound morphemes constituting the grammatical architecture around the lexical stem or kernel are the product of transformations, derivations, and abstractions produced in a morphogenetic process during linguistic change. A standard example is the marking of cases in languages with inflection and their functional parallels in other languages. Four morphogenetic principles can be detected in this field:

---

of the collective, although unconscious, representation, which he called "archetypes" (a term from Greek philosophy). He assumed that these (cultural) archetypes cannot be explained by cultural heritage or geographical diffusion. Instead, they are rather abstract, geometrical constellations that reappear (in Jung's interpretation independently of one another) in many cultures and individuals (see Jung [35]: 45).

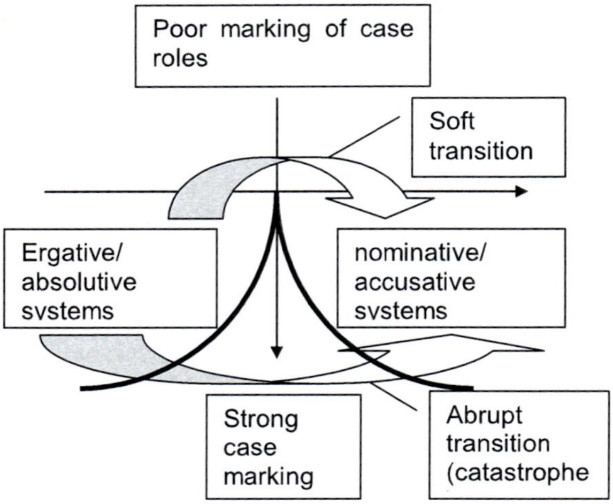

Fig. 6.10 The transition between the ergative and nominative system and the "cusp" catastrophe as the morphogenetic background of this process

(1) Two significant types of case systems emerge from research in comparative linguistics. Systems with a pair of cases called *ergative* and *absolutive* and a pair of cases called *nominative* and *accusative.* In systems with ergative and absolutive cases, the agent in the transitive sentence, which distinguishes an agent from a patient, is marked by the ergative case. The agent in the intransitive sentence (*John* runs) and the patient in the transitive sentence (John throws a *stone*) are marked by the same case category: the absolutive.[20] These languages put their dynamic focus on the agency in the transitive sentence, i.e., the effect of an agent on a patient. In systems of the nominative/accusative type, the patient of the transitive sentence is marked by the case called accusative. In contrast, the agent in both sentences is marked by the case called nominative. In English, this distinction is shown in the pronouns: *He* runs, she kisses him, and in the order of constituents (the case opposition is: he/she versus him/her). Historical linguists discussed whether the two systems have a historical relationship so that marking has transitioned. Such a transition could point to an underlying morphogenetic process with a bifurcation schema. The choice consists of a pole on the scale of the agency. It requires a second term: the patient. A neutral position is a simple type of movement/change. Figure 6.10 illustrates this polarity. The historical change would then be a catastrophic transition that surfaces in languages with a strong marking, for example, by a system of case suffixes or adpositions (see the arrow at the bottom).

(2) In the case of a strong marking, appears either a case system of the type: ergative/absolutive (on the left) or nominative/accusative (on the right). The historical transition takes the form of a bimodal catastrophe. Strong marking can

[20] For simplicity, we do not cite original sentences in languages with an ergative/absolutive case system but the translation of such sentences into English, a language with a nominative/accusative case system.

traverse a more or less abrupt line of change, while the transition is smooth for a weak (or minimal) marking system. This (smooth) transition zone can be observed in language contact, where the case marking is lost. In pidgin languages, a weak or minimal system uses context and contextual knowledge to compensate for the lack of information coded by case marking or adpositions.

Grammaticalization theories assume a gradient with several steps (see Butt [12]: 179): *relational noun > secondary adposition > primary adposition > morphological case affix*. The languages of Indo-European origin show historical transformations leading to a total or partial disappearance of inflectional marking. French and English only mark pronouns, but they have elaborate systems of prepositions. German marks cases in certain noun forms (in the singular or according to noun classes) and shows a replacement of case marking by a marking based on prepositions.

(3) The location of the markers depends on the order of the nominal groups (syntagmas) in a sentence: Subject (S), Verb (V), Object (O); see the simple sentence in German and English: Hans (S) wirft (V) den Stein (O)—John (S) throws (V) the stone (O). The whole series of combinations can be found in the languages of the world: SOV, VSO, VOS, OSV, and OVS. Beyond this typological classification, there exist significant statistical differences. Thus, the order that puts the object in the first position is sporadic; the dominant opposition is that which puts the subject or the verb in the first position.

(4) Local cases appear especially in languages with an extensive list of cases, such as Lesgi (South Dagestan, Caucasus), which has 14 local cases (see Haspelmath [31]). Case localism is generally pertinent, but the agent's field of forces and intentions seems to be the overriding factor. One can call it second-order localism.

Linguistic variation and change have been the central research area of comparative linguistics since its rise in the nineteenth century. Grammatical dynamics were discussed under the concept of "grammaticalization" by Antoine Meillet in 1912. Today we speak of a cycle of grammaticalization based on a scale that goes from lexical entities (especially with concrete, spatial, and imaginal content) to dependent forms to adpositions (prepositions and postpositions) and finally to endings and inflectional paradigms. If the inflectional system disappears, the dynamic restarts at the zero point, i.e., the linear scale is transformed into a cycle. However, cycles of grammaticalization often show transposed phases such that several stages coexist. Therefore, a language can simultaneously have inflectional cases and a set of prepositions that realize case roles (e.g., German or other Indo-European languages have preserved grammatical patterns of the Proto-Indo-European and have replaced many instances with prepositions).

Comparative research has found several routes in the morphogenesis of case markers. They can be derived from a chosen inventory of verbs (especially verbs of movement and spatial change, often in constructions with serial verbs), nouns (often relational), or adverbs. Depending on the typological characteristics of the languages or language families, the dynamics can lead to prepositions or postpositions, suffixes attached to the dependent noun (sometimes also to the verb), or dependent pronouns; see Blake ([9]: 170). Grammaticalization gradients function as morphogenetic fields with categorical transitions between:

- Free lexical units (verb, noun, adverb),
  ↓
- Linked pronouns, adpositions,
  ↓
- Suffixes of the noun (more rarely of the verb),
  ↓
- Inflectional paradigms.

The morphogenetic modeling in grammar can either consider long-term historical (at the limit evolutionary) changes or analyze actual and synchronic processes occurring in natural languages, i.e., either in actual use or short-term developments accessible to observation and analysis.

6.6 Morphogenetic Structures in the Syntax of Verbal Phrases and Sentences

In the tradition of Latin grammar, lexical items and morphology were the core. However, Humboldt, inspired by his study of polysynthetic languages, demanded that the grammar analysis should start from the sentence and not from the word (see his work on the language Kawi in Java, 1830–1835; Humbolt [34]). The psychologist and philosopher Karl Bühler proposed, in 1933, a compromise in the quarrel of contemporary linguists on this question. He postulated that any language knows (at least) two levels of conventionality, distinguishing it from a code of marine signals. These two levels form a scale. At one end, the world is almost torn to shreds, separated into isolated aspects so that each "piece" is associated with a (conventional) sign. On the other end, language tends to per-construct the same world with the help of relations ("Durchkonstruktion") and to establish meaningful gestalts in this construction process (see Bühler [11]: 88). Bühler introduces two levels of construction, segmentation versus construction.

In Chomsky's linguistics, the intermediate notion of the word has been abandoned, and the notion of the sentence has become the key notion of grammar. Fillmore and Kay show in their "Construction Grammar" (cf. Fillmore et al. [23], Goldberg [28]) that the results of generative grammar are easily integrated within a more flexible conception that starts from the notion of construction without abolishing the difference between morphological and syntactic constructions. (Cf. Wildgen [76]: this chapter, pp. 143–169 for an overview in German).

6.6.1 The Morphogenetic Foundation of "Deep" Structures

The distinction between the *deep* form, the force (the "energeia"), and the static *product* was introduced in linguistics at the beginning of the nineteenth century by W. von Humboldt. Chomsky's distinction between deep structure and surface structure, which he abandoned after 1968, was only a technical reflection of this theoretical distinction. It was René Thom ([61]: 121) who reminded us of the deeper motivation of this distinction:

> We, therefore, admit that the hypothesis of a 'deep structure' of linguists consists essentially of our sensory representation of the external world (barely elaborated by perception). On the contrary, the surface structure will be constituted by automatisms of language themselves; they constitute a layer of spaces superficially attached to the "deep structure", and historically, in evolution, they come from it by a process of permanent exfoliation, like our skin, made up of layers of cells secreted by the deep dermis and which go in the process of sclerosis towards the outside, where they disintegrate. (Translation by the author)[21]

René Thom's biological interpretation of Humboldt's notion of "energeia" in grammar and the traditional notion of "deep" or crypto-structure in linguistics presupposes a very general concept of "morphology", as Goethe coined it in his "Morphologie überhaupt". It links the biological forms in morphogenesis to symbolic forms, such as language. Thom [61] argued that linguistics is an exemplary morphological discipline. This means that the reference to biological morphogenesis is not just a superficial analogy; morphological principles visible in biology are concentrated in symbolic forms, specifically in language. Linguistics is the prototype of morphology.[22]

6.6.2 Semantic Roles and the Dynamics of Sentential Frames

The intuition that sentences and verbal phrases operate with a finite and small set of schemes or construction types goes back to antiquity. Wildgen ([68]: Chap. 2, pp. 9–58; in German) discussed this tradition up to modern case theories. The morphogenetic approach proposes a radically biological/cognitive analysis. It is founded in the publications of René Thom, who formulated the initial conjecture and elaborated in Wildgen ([69]; short English version in Wildgen [66]). A detailed analysis was published in Wildgen [72], in English, Chaps. 3 and 5) and in Wildgen

[21] "Nous admettons donc que l'hypothétique d'une 'structure profonde' des linguistes est constituée essentiellement de notre représentation sensorielle du monde extérieur (à peine élaborée par la perception). Au contraire, la structure de surface sera constituée par des automatismes du langage proprement dits; ils constituent une couche d'espaces accolés superficiellement à la "structure profonde", et historiquement, dans l'évolution, ils en proviennent par un processus d'exfoliation permanente, à la manière de notre peau, constituée de couches de cellules secrétées par le derme profond et qui vont se sclérosant vers l'extérieur, où elles se désagrègent."

[22] The relation between Goethe's concept of "Morphologie überhaupt" and Thom's morphogenetic access to language is discussed in Wildgen ([67]; in German).

([73]; in French, First Part: "La grammaire morphodynamique"). The account in the present book only considers the general lines and principles. Two assumptions are characteristic:

- Semantic roles are primarily a cognitive classification of processes (states) subject to verbalization. Therefore, their projection in a morphological and syntactic classification is polysemic and lacunar, i.e., the role must not be expressed in every context and can be left to the listener's interpretation.
- Semantic roles are variables and depend on the phenomenological level in which the process (see the list below) is rooted.

We consider a hierarchy of levels centered on the individual speaker/hearer. The topological proximity to the ego (from the periphery to the center) motivates the following process levels (the stable state is considered the limiting case of a process):

(a) Processes in the ambient space of the speaker/actor,

- *local* processes; the processes take place in the sphere of the subject, for instance, bodily movements of the agent.
- *interlocal* processes; the support of the process, for instance, a subject or an object changes his/its location.

(b) Sensory processes: The center of the process is located in the (peripheral) apparatus of the senses.
(c) Mental (self-referential) processes: The center is the brain (cortex).
(d) Qualitative/quantitative changes: They are quasi-external to the individual but depend on quality dimensions and quantitative scales that the individual has internalized (often through language and cultural techniques).
(e) Abstract changes: They constitute a heterogeneous class. Their meaning is vaguer, and the referential source remains opaque.

The system of levels (a)–(e) can be represented as a system of containers (circles) around and inside the space centered on the Ego (Fig. 6.11).

The scene (the drama) described by a simple sentence (with a finite verb) is broken down into several regimes (sub-centers), which we call processual (semantic) roles. The verb represents the type of process. The classification of possible scenes gives a system of representations we call imaginal ("imaginistic" according to Kosslyn [40]). They are neither images nor perceptual structures but cognitive entities at a syncretic level on which the grammatical morphogenesis can operate.

The fundamental question that René Thom asked and to which he was able to give a surprising answer is the following: The scenes are continuous and contain an unlimited number of variables that can influence what happens. Is there a possibility of finding a finite list of stable patterns to which all these variations can be reduced? His solution has been elaborated regarding linguistic facts and neuro-cognitive research; cf. for the first direction Wildgen [66, 72, 81] and the second Petitot [49, 50].

As sketched in Thom [60], the morphology of sentential expressions points to morphogenesis rooted in fundamental biological and social behavior like grasp,

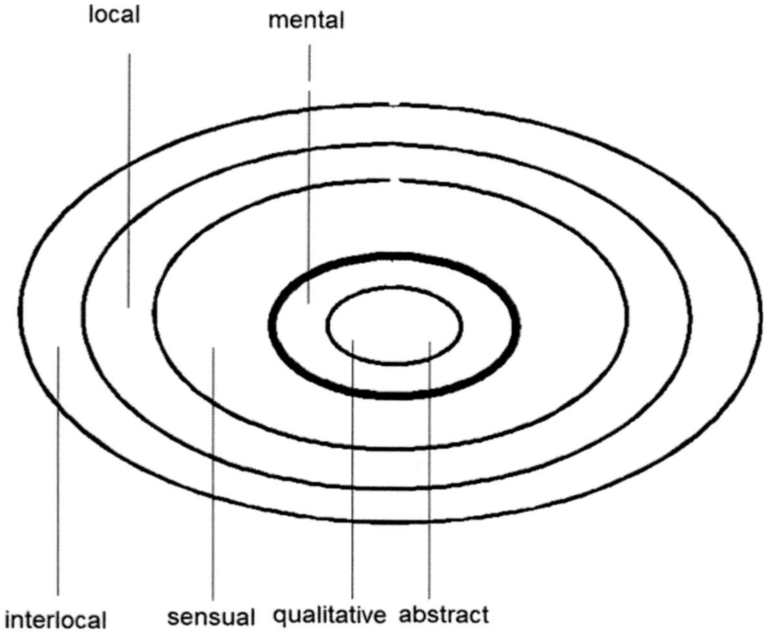

Fig. 6.11 The phenomenological levels of process scenarios

predate, gift, and media of exchange (commercial and symbolic). Basic invariants of these processes are analog to a classification of the stable unfolding of dynamical systems in differential topology (specifically in the results of catastrophe theory, further elaborated in bifurcation analysis and chaos theory). The realization of these schemata in the grammars of current or historically documented grammars is polysemic and lacunar. Still, the empirical evidence supports Thom's hypothesis (with a marge of statistical insecurity). As no better explanation is available (beyond a pure description of superficial evidence), his proposals remain relevant until better, biologically rooted theories come to the fore.

6.7 Morphogenetic Patterns in the Syntax of Nouns and Adjectives

The noun phrase is centered (usually) on a nominal nucleus (a noun) connected to a certain number of satellite words. According to Thom (1972), this connection is the product of a fundamental operation that appears in biological evolution and reappears in man's cognitive development. We can consider four possible sources of this relationship:

(1) Aristotle had already noticed that the hierarchy of the genera in biology corresponds to the lexical hierarchy of classificatory labels. The "species" is distinguished from the "genus proximum" by a "differentia specifica".
Example: man (genus) is a social (specification) animal (species).

(2) The parts of the human body are cognized as a tree-like hierarchy (an acyclic graph). The different parts of the head can be hierarchized according to the way indicated in Fig. 6.12 (see Thom [63]: 187). Languages differ in the way they achieve the taxonomy of body parts.
The denominations of body parts, as well as those of any other hierarchies like that of kinship terms, are called relational because they always imply a whole network or a local neighborhood of this network:

- the forehead (of the skull),
- the mother (of a child).

The general idea that emerges from these systems is called: "diffusion de prégnance" (spread of relevance). If this diffusion is acyclic, it automatically leads to tree-like structures.

(3) The meaning of an adjective or some specifying attribute of the noun can be a secondary index. Already in animals, such an indexical mark can be the trace, the excretum of the predator that is perceived in its absence. In this case, an alarm call negates the predator's presence but tells its former (and possibly future) presence. In a nominal construction, the predator would be the noun, and the trace or excretum the attribute. In the context of religion, the attribute (epithet) is a sign for a god, for instance, thunder for the god Zeus. Thom ([63]: 28) argues that the trace is a typical morphogenetic source of attributes (epithets): In the case of a genitive construction in the noun phrase, it is derived from the noun by case marking.

"The 'genitive', a syntactical form which indicates the proximity of a being but at the same time denies its immediate presence, may have appeared to resolve this dilemma; at the sight of a trace, one raised the cry of alarm but affixing it with an affix that negates the actual presence of the predator, allowing for a

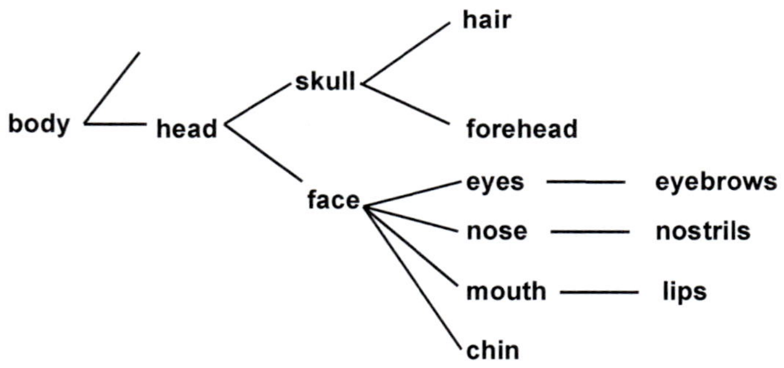

Fig. 6.12 Tree-like hierarchy of body (head) parts in French (translated to English by the author)

more graded form of defense strategies."[23] In more general terms, the adjective, attribute, or ad-nominal modifier separates a qualification from its source, the substratum, in the process of abstraction.

(4) Children's drawings show an association of parts in the figure that hardly respect correct geometric and metric relations (measures). In these drawings, it is natural to see a trace of the conceptual structure acquired by the child, where parts and aspects are separated and then loosely associated with the whole figure. For example, Thom writes ([64]:179):

However, when we put a concept in the genitive, we dissociate it into its fundamental elements, [...] that is to say, we highlight all the sub-concepts which intervene in the meaning, that is, the regulation of the mental figure of the concept: the tail of the dog, the wheel of the car, etc.[24]

The different aspects mentioned by René Thom open the way to a (cognitive, even a biological) explanation of the adjective and the syntactic construction of nominal phrases.

6.7.1 The Positional Hierarchies of the Adjective and Its Semantic Values

The analysis by Hansjakob Seiler [54] will be our starting point. Seiler's model considers a continuum of determination with two extremes:

- The specifications of the referential relation (extension). Typical realizations are deictic gestures and demonstratives.
- Descriptive characterizations (intension).

The German sentence (translated word by word) shows the typical syntactic organization:

alle diese meine/erwähnten zehn schönen roten hölzernen Kugeln
all-those-my/mentioned ten beautiful red wooden balls
des Spiels auf dem Tisch, die ich dir jetzt gebe, ein Geschenk
of the game on the table that I give you now as a gift ...(apposition).

The central noun is "Kugeln" (balls). It is preceded by two groups of determinatives (in the broad sense):

[23] «Le 'génitif', forme syntaxique qui indique la proximité d'un être mais en même temps nie sa présence immédiate, est peut-être apparu pour résoudre ce dilemme; à la vue d'une trace on a poussé le cri d'alarme mais en l'affectant d'un affixe qui niait la présence effective du prédateur, ce qui permettait une forme plus graduée des stratégies de défense.»

[24] «Or, quand on met un concept au génitif, on le dissocie en ses éléments fondamentaux, [...], c'est-à-dire qu'on met en évidence l'ensemble des sous-concepts qui interviennent dans la signification, c'est-à-dire la régulation de la figure mentale du concept: la queue du chien, la roue de la voiture etc.»".

(a) "alle diese meine",
(b) "erwähnten zehn schönen roten hölzernen".

Seiler calls the border between (a) and (b) the turning or inflection point. While in (a) the order is rigid, it is flexible in (b) but governed by semantic laws. The order of epithetic adjectives responds to a criterion of specificity. Adjectives closer to the noun (relative to others) also have a more natural link to the center; they are more specific for that noun. In German, the order of categories corresponds to the following scale:

numerals (1), evaluative adjectives (2), color adjectives (3), adjectives of substance (4), and noun nucleus (5) (cf. for details Wildgen, 1999: 211–214; in French).

The descriptive space of noun and adjective phrases is characterized by its semantic or informational complexity; cf. [62]: 81). For example, the proper name "René Thom" is semantically more complex than the nouns "mathematician", "professor", "French", and "man" because by forgetting specific characteristics of the individual René Thom, we come to the mathematician, the professor, the French, and the man. The maximum value is given by a complete description that specifies for all predicates whether the subject fulfills them or not. Disjunctions (choices) instead of conjunctions diminish the informational value,finally, if none of the predicates is fixed in its truth value, the information is zero (cf. for the notion of semantic information Carnap and Har Hillel, 1952 and for its application Wildgen [77].

In general, it can be said that the noun phrase has its specific syntactic (morphological) and semantic laws. It refers to a semantic continuum with categorization (catastrophe) points, categorical focuses (attractors), and gradients on this continuum.

6.7.2 *Sketch of the Morphogenetic Structure of Noun Phrases*

Suppose we start from a continuum on which regions are defined around a focus (the categories of determinants and the nominal kernel). Then, we can choose a dynamical system with a dominant attractor and several attracting satellites as a basic model. Figure 6.13 shows us the graph of a potential function, and below, the Dynkin diagram retains only the critical points: maxima (–) and minima (+).

NN designates the noun category; typical members of this category are words that designate stable entities existing in the neighborhood of humans or their imagination rooted in this environment (for instance, discrete entities around the speaker). Mass nouns (water, milk, steam) and abstract nouns (whiteness, virtue, and happiness) are at a certain distance from the center of the prototype of the nominal category. We can predict that the satellites' categorical weight (nominality) decreases with the distance from the central attractor. This distance from the prototype of the category noun ("nominality") defines what we call the "categorical distance" in the noun phrase.

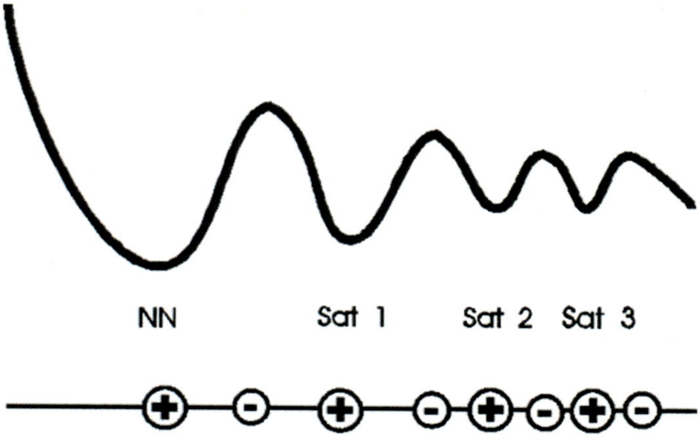

Fig. 6.13 Dynamic and discrete representation of a nominal kernel (NN) with its satellite concepts (Sat1, 2, 3)

Additionally, one has to consider a second dimension called information complexity. A third dimension is given by the context and the momentary state of the speaker and his interlocutor. In a nominal group, we can distinguish elements that refer to the situation from elements that contribute information. For example, the deictic determinatives (this, this one), pronouns (I, you, my, your), articles (one, the), and quantifiers (no, five, all) refer to the situation. On the other hand, adjectives, nominal attributes, participles, and relative clauses contribute to the information in the nominal phrase, i.e., to the second dimension. Cf. Wildgen [77]: 218f) for more details on this spatial reconstruction of the semantics of noun phrases (in French).

Nominal constructions must provide positions that fill the dimension: information and contextual (indexical) rooting and provide enough distinctions. Languages differ in how they categorize this semantic space of noun phrases. What is universal is not a set of grammatical categories and subcategories but the space itself and the inherent dynamics of differentiation.[25]

The morphogenetic model allows us to predict the most frequent types of differentiation under the assumption that stable differentiations have maximally three or (under special conditions) four attractors (cf. the restrictions on valences treated in Sect. 6.6). Sequences on the same level can be iterated without theoretical limits but confined by limits of memory or attention.[26]

[25] It was a flaw of typological and comparative research in linguistics that it tried to use the traditional labels introduced in the grammar of classical Greek or Latin in the analysis of a corpus of different languages. Even if this list is enlarged or modified, the problem remains that a list choosing ad hoc between distinctions found in specific languages cannot be a general background of linguistic analysis. In any human language, the selection of "deep cases" and their expression results from an epigenetic process depending on the contingent, i.e., aleatory factors.

[26] Chomsky compared this iteration to complete algebraic induction and thus argued for an algebraic modelling in syntax. In reality, the phenomenon of recursiveness in natural languages is rather due to

The construction of a low-dimensional semantic space is only a starting point. First, any theorization must pass through the stage of ideal construction. At this level, morphogenesis and possible biological or cognitive determinants may be treated. Beyond this basic level, the aleatory nature of symbolic systems (see the "arbitraire du signe" in de Saussure's "sémiologie") comes to the fore, and questions of explanatory relevance become opaque or even inaccessible.

6.8 Morphogenesis on Different Scales and the Stability of Language (and Other Symbolic Forms)

The morphogenetic perspective on language described in this chapter has highlighted several layers of semiosis that must be considered. These results can be extrapolated to the morphogenesis of the symbolic forms analyzed in Chaps. 3, 4 and 5. Moreover, it seems necessary to get a picture of the relevance of morphogenesis for humans and human societies as wholes. The list of layers starts from the oldest ones, those that have governed the history of humankind for many millennia:

(a) The emergence of a phonic language with a systematic impact on world knowledge and practical control of the ambient sphere. This morphogenetic process has separated the human species (Homo sapiens) not only from its predecessors, for instance, chimpanzees, the last common ancestor of chimpanzees and humans (LCA), but also from Homo erectus who expanded from Africa to many parts of the world, Homo heidelbergensis and finally from the subspecies of Neanderthals and Denisovans that have left a genetic trace in the genome of many human populations. It remains a controversial issue if the Neanderthals and the Denisovans had linguistic capacities comparable to those of ancient Homo sapiens or even to actual human populations. The morphogenesis of language has been modeled in this book using scenarios of (iterated) bifurcation and (tentatively) the consideration of hypercycles in the sense of [22]. Such highly competitive and selective processes could explain this evolution's speed and quasi-goal-directed nature.

(b) Beyond the pure existence of a phonic language with the capacity to code referential meaning, the morphogenetic analysis of language must explain the organization of a rich lexicon and the emergence of techniques for the composition of meanings in morphology, syntax, and discourse, i.e., the richness of human languages in all known societies. Beyond a primitive lexicon (not far from lists of referential cues in animals, i.e., 10 to 100 items) and a "syntax" of juxtaposition comparable to two-word utterances in early childhood or primitive (ad hoc) pidgins, we assumed a process of self-organization that reduces the insecurity/instability of a syntax of juxtaposition. The transition was only

the lack of topological constraints. Semantically an infinite series of attributes has a chaotic attractor; the meaning of the nominal construction is annihilated. Cf. my remarks on the accumulation of attributes in the characterization of God in Sect. 5.8.2.

possible under the assumption of a precise and rapid production and memory of phonic patterns (syllables, words, phrases) and an efficient reorganization of the growing lexicon that is analogous to human mental capacities of spatial control in locomotion, handling of objects, and fundamental interactions with other animate beings. The key to such semantics of phonic utterances is constructing a semantic space stabilized by its roots in a low-dimensional semantic space (basically three dimensions). The classification theorem of catastrophe theory can explain the generality of such a restriction valid for simple structurally stable dynamical systems and their unfolding in time. The schematizations of processes, scenarios of events, and actions on this basis are restricted in the elementary case to three dimensions, and under special conditions to a fourth dimension. Beyond these limits, the stability of meaning constructions (semiosis) is endangered, and specific measures must be taken to avoid chaotic or even aleatory effects. The technique of such a reduction for the sake of stability can be observed in the analysis of case systems and similar syntactic devices (this generalization is caught under the term "deep cases" and was a central concern of grammars already in antiquity). In Sect. 6.6, some results in catastrophe theoretic semantics were summarized (avoiding the technicality of a formal model).

(c) Schematizations in the lexicon of verbs (and other relational lexical items) and sentential constructions are the third manifestation of morphogenesis in language. They must have an image-like character (a quasi-spatiality). This means that scenarios of real life with a high degree of relevance ("prégnance" in terms of René [63] must be coded in every grammar of a human language such that despite the temporal/sequential mode of phonic language, a quasi-spatial meaning can be recovered by the addressee of the utterance. This technique is the key to effective communication and opens the door for efficiency in everyday practices like language, art, music, religion, and other symbolic forms. The rich epistemic systems characteristic of human cultures in science and other symbolic systems enable humans to grasp, manage, and control their environment (including the personal and socio-political sphere).

Chapters 3, 4 and 5 show that the symbolic forms of music, art, and religion responded to building highly organized spaces of meaning that allow for creativity and the establishment of prosperous and stable traditions. These are the necessary frames for communication and innovation, whereby the traditions are consistently adapted to changing conditions. Innovations may even enforce a radical change in the traditions, mainly if these are not appropriately adapted to new conditions. Beyond such a breaking of tradition, often accompanied by destructive decisions and social conflict, two basic demands must be fulfilled: First, an amount of successful understanding between community members must be guaranteed; second, a minimal level of reliability ("reality") must be sustained. Human populations cannot survive in an illusionary or fake "reality".

Suppose we venture into a global (historical and geographical) perspective. In that case, we may consider the past of humanity and possibly its future as a morphogenetic process in which individual decisions, even those of leaders and people with a maximal concentration of influence and power, are not decisive (their effect is just a kind of small-scale variation). The history of humankind and even the time in which the earth is habitable are only insignificant spots in astrophysical processes. The maximal range of individual decisions does not exceed two or three generations; societies and empires may exist for centuries (the range of the Roman empire covers fewer than 2000 years). Ratiogenetic processes like human planning and politics have a much shorter range. What remains relatively constant is the species that may stay more or less identical for 100,000 or even a million years. In the same period, innumerable other processes occur in parallel and dramatically change the conditions of survival and subsistence.

In Chap. 5 on myth and religion, the visions of the end times and the apocalypse have been discussed. Currently, many intellectual activities concern the future of the climate on earth, the danger of global diseases, and possible scenarios of a third world war. However, such projections into the future of humanity and possible interventions to influence this evolution must consider the diversity of morphogenetic and self-organizing processes beyond the influence of human agents.

The symbolic forms considered in this book, i.e., music, art, religion, and language (possibly also further ones like ethics, technologies, and science), have a common feature: They have a kind of autonomy concerning physical, economic, and political controls. Moreover, they belong to the realm of common goods[27] accessible to everybody, like the air we respire, the water we drink, and the ground we are moving on. In this sense, they cannot be subdued to the interests of single persons or social groups. Therefore, a kind of natural evolution in concert with other natural processes cannot be suppressed or evinced. Consequently, the impact of human caprice or despotism is limited. However, this does not mean that natural processes secure the future of humanity. On the contrary, humanity may naturally disappear or destroy itself.

References

1. Allott, R.: The origin of language: the general problem. In: Wind, J., Pulleyblank, E., de Groher, E. and Bichakjian B, H. (eds.) Studies in Language Origins, vol. 1. pp. 1–24. Benjamins, Amsterdam (1989)
2. Allott, R.: The motor theory of language. In: Raffler-Engel, W., Wind J., Jonger, A. (eds.) Studies in Language Origins, vol. 3, pp. 123–157 (1991)
3. Allott, R.: Motor theory of language origin: the diversity of languages. In: Wind, J., Jonker, A., Allott, R., Rolfe L. (eds.): Studies in Language Origins, vol. 4. pp. 125–160. Benjamins, Amsterdam (1994)
4. Aristotle: de Anima. Clarendon Press (ed. Shields, Ch.J.). Oxford U.P., Oxford (2016)

[27] This notion was already present in the work of Aristotle. Rousseau, in his book "The Social Contract", 1762), and Adam Smith, in his book "The Wealth of Nations", 1776), have elaborated the notion in terms of modern politics and economics..

5. Ashby, W.R.: Principles of the self-organizing dynamic system. J. Gen. Psychol. **37**(2), 125–128 (1947)
6. Ballmer, T.T., Brennenstuhl, W.: Deutsche Verben. Eine sprachanalytische Untersuchung des Deutschen Wortschatzes. Narr, Tübingen (1986)
7. Berlin, B.: Ethnobiological classification. In: Rosch, B., Lloyd, B.B. (eds.) Cognition and Categorization. Wiley, New York (1978)
8. Blackmore, S.: The Meme Machine. Oxford U.P, New York (1999)
9. Blake, B.J.: Case. Cambridge U.P., Cambridge (2001)
10. Boesch, Ch: Aspects of transmission of tool-use in wild chimpanzees. In: Gibson, K.R., Ingold, T. (eds.) Tools. Language and Cognition in Human Evolution, pp. 171–183. Cambridge U.P, Cambridge (1993)
11. Bühler, K.: Sprachtheorie. Die Darstellungsfunktion der Sprache, 2nd edn. Fischer, Stuttgart (1965)
12. Butt, M.: Theories of Case. Cambridge U.P, New York (2006)
13. Cangelosi, A., Greco, A., Harnad, St.: Symbol grounding and the symbolic theft hypothesis. In: Cangelosi A., Parisi, D. (eds)Simulating the Evolution of Language, pp. 191–210. Springer, Leipzig (2002)
14. Chomsky, N.: Syntactic Structures. Mouton, The Hague (1957)
15. Chomsky, N.: Cartesian linguistics. Harper & Row, New York (1966)
16. Chomsky, N.: Some simple evo devo yheses: how true might they be for language? In: Larson, R.K., Déprez, V., Yamakido, H. (eds.) The Evolution of Human Language, pp. 45–62. Biolinguistic Perspectives. Cambridge U.P, Cambridge (2010)
17. Comrie, B., Kuteva, T.: The evolution of grammatical structures and 'functional need' explanations. In: Tallermann, M. (ed.): Language Origins: Perspectives on Evolution, pp. 185–207. Oxford U.P., Oxford (2005)
18. Corballis, M.C.: The Evolution of Language. Annals of the New York Academy of Science (2009) https://doi.org/10.1111/j.1749-6632.2009.04423.x
19. Dediu, D., Lewinson, St.: The time frame of the emergence of modern language and its implications. In: Dor, D., Knight, Ch., Lewis J. (eds.), The Social Origins of Language, pp. 184–195. Oxford U.P., Oxford (2014)
20. Dor, D., Knight Ch., Lewis, J.: The Social Origins of Language. Oxford U.P., Oxford (2014)
21. Dunbar, R.: Groups, Gossip, and the Evolution of Language. In: Schmitt e.a. (eds.): New Aspects of Humans Ethology. Plenum Press, New York, pp. 77–89 (1997)
22. Eigen, M., Schuster, P.: The Hypercycle: A Principle of Natural Self-organization. Springer, Berlin (1979)
23. Fillmore, Ch., Kay, P., and O'Connor, C.: Regularity and Idiomaticity in Grammatical Constructions. The Case of Let Alone. Language 64, 501–538 (1988)
24. Fisher, S.E.: Evolution of language: lessons from the genome. Psychon. Bull. Rev. **24**, 34–40 (2017). https://doi.org/10.3758/s13423-016-1112-8
25. Fitch, T.W.: The Evolution of Language. Cambridge U.P, Cambridge (2010)
26. Gibson, J.J.: The Ecological Approach to Visual Perception. Houghton Mifflin, Boston (1979)
27. Gil, D.: Early Human Language was Isolating-Monocategorical-Associational (2006) http://www3.isrl.uiuc.edu/~junwang4/langev/localcopy/pdf/gil06evolang.pdf
28. Goldberg, A. E.: Constructions: A Construction Grammar Approach to Argument Structure. Chicago U.P., Chicago (1995)
29. Haken, H.: Synergetics. An Introduction (third revised and augmented edition). Springer, Berlin (1983)
30. Haken, H.: Principles of Brain Functioning: A Synergetic Approach to Brain Activity, Behavior, and Cognition. Springer, Berlin (1996)
31. Haspelmath, M.: Grammar of Lezgian. deGruyter, Berlin (1993)
32. Hauser, M.D., Chomsky, N.: Fitch: the language faculty: what is it, who has it, and how did it evolve? Science **89**, 1569–1579 (2002)
33. Henke, K., Reber, T.P., Duss, S.B.: Integrating events across levels of consciousness. Conscious Cogn. **21**(2), 953–960 (2012). https://doi.org/10.1016/j.concog.2012.02.013. Epub 2012 Mar 17

34. von Humboldt, W.: Über die Kawi-Sprache auf der Insel Java : nebst einer Einleitung über die Verschiedenheit des menschlichen Sprachbaues und ihren Einfluss auf die geistige Entwickelung des Menschengeschlechts. Dümmler, Berlin (1839)
35. Jung, C.G.: Die Archetypen und das kollektive Unbewußte. Gesammelte Werke, vol. 9 (1). Rascher, Zürich (1984)
36. Katz, J., Fodor, J.: The structure of a semantic theory. Language **39**, 170–210 (1963)
37. Kelso, J.A.S.: Dynamic Patterns: The Self-organization of Brain and Behavior. MIT Press, Cambridge (Mass.) (1995)
38. Kendon, A.: The 'polymodalic' nature of utterances and its relevance for inquiring into language origins. In: Dor, Knight, Lewis, pp. 67–76 (2014)
39. Kirby, S.: Syntax without natural selection: how compositionality emerges from vocabulary in a population of learners. In: Knight, Ch., Studdert-Kennedy, J M., Hurford, R. (eds.): The Evolutionary Emergence of Language. Social Functions and the Origins of Linguistic Form, pp. 303–323. Cambridge U.P., Cambridge (2000)
40. Kosslyn, S.M.: Image and Mind. Harvard U.P, Cambridge MA (1980)
41. Labov, W.: The Boundaries of words and their meanings. In: Fishman, J. (ed.) New Ways of Analysing Variation in English, pp. 340–373. Georgetown U.P., Washington D.C. (1973)
42. Lakoff, G., Johnson, M.: Metaphors We Live By. Chicago U.P, Chicago (1980)
43. Levinson, S.C.: Space, Language, and Cognition: Explorations in Cognitive Diversity. Cambridge U.P., Cambridge (Mass.) (2001)
44. Lewis, J.: BaYaka pygmy multimodal and mimetic communication traditions. In: Dor, Knight, and Lewis, pp. 77–91 (2014)
45. McNeill, D.: Hand and Mind: What Gestures Reveal About Thought. Chicago U.P, Chicago (1992)
46. Orban, G.A.: Neuronal Operations in the Visual Cortex. Springer, Berlin (1984)
47. Osgood, Ch. E., Suci, G.J., Tannenbaum, P.H.: The Measurement of Meaning. Urbana (1957)
48. Petitot, J.: Morphogenèse du sens, Presses Universitaires de France, Paris. (1985) [English translation: Morphogenesis of Meaning. Lang, Bern (2004)]
49. Petitot, J.: Physique du sens. De la théorie des singularités aux structures sémio-narratives. Éditions du CNRS, Paris (1992)
50. Petitot, J.: Cognitive Morphodynamics: Dynamical Morphological Models of Constituency in Perception and Syntax. Peter Lang, Bern (2011)
51. Petitot, J.: Neurogéométrie de la vision. Les Éditions de l'École Polytechnique, Paris (2008) English translation : Elements of Neurogeometry: Functional Architectures of Vision (Lecture Notes in Morphogenesis). Cham, Springer (2018)
52. Rizzolatti, G., Arbib, M.A.: Language within our grasp. Trends Neurosci. **21**(5), 188–194 (1998)
53. Rosch, E.: Human categorization. In: Warren, N. (ed.): Studies in Crosscultural Psychology 1, pp. 3–50. Academic Press, New York (1977)
54. Seiler, H.: La dynamique dans la dimension de la possessivité. In Petitot, J. (ed.): Logos et théorie des catastrophes. À partir de l'œuvre de René Thom, pp. 409–418. Geneva, Editions Patiño (1988)
55. Starck, D.: Stammesgeschichtliche Voraussetzungen der Entwicklung der menschlichen Sprache. Nova Acta Leopoldina N.V. 54 (245), 581–596 (1981)
56. Steels, L.: Grounding symbols through evolutionary language games. In: Cangelosi, A., Parisi, D. (eds.) Simulating the Evolution of Language, pp. 211–226. Springer, London (2002)
57. Steels, L.: Breaking down False Barriers to Understanding. In: Dor, D., Knight Ch., Lewis, J. (eds) The Social Origins of Language, pp. 336–339. Oxford U.P., Oxford (2014)
58. Studdert-Kennedy, M.: Evolutionary implications of the particulate principle: imitation and the dissociation of phonetic form from semantic function. In: Knight, Ch., Studdert-Kennedy, J M., Hurford, R. (eds.) The Evolutionary Emergence of Language. Social Functions and the Origins of Linguistic Form, pp. 161–176. Cambridge U.P., Cambridge (2000)
59. Timmermann, A. et al.: Climate effects on archaic human habitats and species successions. Nature **604** (2021)

60. Thom, R.: Stabilité structurelle et morphogenèse, Interéditions, Paris (English translation: Structural stability and morphogenesis. Benjamin, Reading, 1975) (1972)
61. Thom, R.: La linguistique, discipline morphologique exemplaire. Critique **30**(1), 235–245 (1974)
62. Thom, R.: La double dimension de la grammaire universelle. In: Seiler H-J. (ed.) Language Universals. Papers from the Conference held at Gummersbach, pp. 79–87. Narr, Tübingen (1978)
63. Thom, R.: Esquisse d'une sémiophysique: Physique aristotélienne et théorie des catastrophes. Interéditions, Paris [English translation: Semiophysics. A Sketch. Addison-Wesley, Boston 1990] (1988)
64. Thom, R.: Apologie du Logos. Hachette, Paris (1990)
65. Wiener, N.: Cybernetics: Or Control and Communication in the Animal and the Machine. Hermann & Cie, Paris & MIT Press, Camb. Mass (1951)
66. Wildgen, W.: Catastrophe Theoretic Semantics: An Elaboration and Application of René Thom's Theory. Benjamins, Amsterdam (1982)
67. Wildgen, W.: Goethe als Wegbereiter einer universalen Morphologie (unter besonderer Berücksichtigung der Sprachform). In: Jahresbericht des Präsidenten 1982, University of Bayreuth, pp. 235–277 (1983)
68. Wildgen, W.: Archetypensemantik. Grundlagen einer dynamischen Semantik auf der Basis der Katastrophentheorie. Narr, Tübingen (1985a)
69. Wildgen, W.: Dynamische Sprach- und Weltauffassungen (in ihrer Entwicklung von der Antike bis zur Gegenwart. Universitätsverlag, Bremen (1985b) https://www.academia.edu/28889869/Dynamische_Sprach_und_Weltauffassungen_in_ihrer_Entwicklung_von_der_Antike_bis_zur_Gegenwart_
70. Wildgen, W.: Das dynamische Paradigma in der Linguistik .[The Dynamical Paradigm in Linguistics], electronic version of part 1 in: Wildgen and Mottron Dynamische Sprachtheorie. Sprachbeschreibung und Spracherklärung nach den Prinzipien der Selbstorganisation und der Morphogenese, Brockmeyer, Bochum. [Dynamic Theory of Language. Linguistic Description and Explanation Following the Principles of Self-Organization and Morphogenesis] (1987) https://elib.suub.uni-bremen.de/ip/docs/00010028.pdf (1987/2005)
71. Wildgen, W.: Basic principles of self-organization in language. In: Haken, H. and Stadler M. (eds.): Synergetics of Cognition, pp. 415–426. Springer, Berlin (1990)
72. Wildgen, W.: Process, Image, and Meaning. A Realistic Model of the Meanings of Sentences and Narrative Text. Benjamins, Amsterdam (1994)
73. Wildgen, W. : De la grammaire au discours. Une approche morphodynamique. Peter Lang, Bern [From Grammar to Discourse. A Morphodynamic Aproach] (1999)
74. Wildgen, W.: The Evolution of Human Languages. Scenarios, Principles, and Cultural Dynamics. Benjamins, Amsterdam (2004a)
75. Wildgen, W.: The paleolithic origins of art, its dynamic and topological aspects, and the transition to writing. In: Bax, van Heusden, Wildgen (eds.) Semiotic Evolution and the Dynamics of Culture, pp.117–153. Lang, Bern (2004b)
76. Wildgen, W.: Kognitive Grammatik: Klassische Paradigmen und neue Perspektiven. de Gruyter, Berlin (2008a)
77. Wildgen, W.: Semiotic hypercycles driving the evolution of language. Axiomathes **18**(1), 91–116 (2008b)
78. Wildgen, W.: Sketch of an evolutionary grammar based on comparative biolinguistics. In: Röska-Hardy, L.S., Neumann-Held, E.M. (eds.) Learning from Animals? Examining the Nature of Human Uniqueness, pp. 45–59. Psychology Press. Hove and New York (2009)
79. Wildgen, W.: Thom's theory of „saillance" and „prégnance" and modern evolutionary linguistics. In: Wildgen, W., Brandt P. Aa. (eds.) Semiosis and Catastrophes. René Thom's Semiotic Heritage, pp. 79–100. Lang, Bern (2010)
80. Wildgen, W.: Language evolution as a cascade of behavioral bifurcations. ELUA **26**, 359–382 (2012)

81. Wildgen, W.: The cultural individuation of human language capacity and the morphogenesis of basic argument-schemata. In: Sarti, A., Montanari, F., Galofaro, F. (eds.) Morphogenesis and Individuation, pp. 93–110. Springer, Berlin (2015)
82. Wildgen, W.: En cas de catastrophe. Les systèmes casuels et la dynamique qualitative. Estudos Semióticos **13**(1), 1–15 (2017). https://www.revistas.usp.br/esse/article/download/138414/133874
83. Zadeh, L.A.: Fuzzy sets. Inf. Control **8**(3), 338–353 (1965) https://web.archive.org/web/20150813153834/http:/www.cs.berkeley.edu/~zadeh/papers/Fuzzy%20Sets-Information%20and%20Control-1965.pdf

Chapter 7
Peirce's Semiotics, Cassirer's Philosophy of Culture, and the Epistemology of Semiotics

Abstract The analysis of symbolic forms asks for an epistemology tuned to its object and an ontological commitment that goes beyond a monist (physicalist) ontology or a dualist ontology, such as Descartes' distinction between matter ("res extensa") and mind ("res cogitans"). Peirce's comment on Kepler and his scientific strategy gives a good starting point. The adequate ontology for semiotics (and symbolic forms) must distinguish three local ontologies and their relations: the world (W), the individual mind (I), and the community of communication (C). The relationships in this triangular constellation define a network of information transfers and specific formats (propositional, spatial, or continuous). Peirce's central notion of "interpretant" is elaborated in this context.

7.1 Introduction

The post-medieval view was that the space of allowed thinking, imagination, and innovation is strictly confined and has to respect some standard interpretation of the Holy Scriptures (under the authority of some acknowledged institution). For Galilee and Kepler, the new heliocentric model was mainly a technical (mathematical) innovation, but, in reality, it triggered a series of intellectual revolutions.[1] Renaissance philosophy opened the field of philosophical and cosmological consciousness beyond medieval restrictions. Giordano Bruno's model of an infinite space of worlds (solar systems) and human conceptualization and memory opened the doorway to European Enlightenment and modern science in the seventeenth and eighteenth centuries (cf. Wildgen [30, 34] for "cosmological memory" in Giordano Bruno's opus in Latin). The eighteenth century made a big step in introducing the anthropological dimension of language and thinking,it was the starting point of all modern explanations of

[1] They thought it would not provoke any conflict with the religious creed. One had only to accept minor modifications related to marginal passages in the Scriptures. However, Giordano Bruno had already grasped the provocative power of the Copernican worldview in 1584; cf. Widlgen [33].

W. Wildgen, *Morphogenesis of Symbolic Forms: Meaning in Music, Art, Religion, and Language*, Lecture Notes in Morphogenesis,
https://doi.org/10.1007/978-3-031-25651-6_7

the origin of human language and human cognition (even before Darwin explained evolution in 1859; cf. the writings of Condillac, Maupertuis, Rousseau, Diderot and Herder after 1746).

However, the infiniteness of the space of consciousness does not mean that in a specific moment, for specific operations, the human mind is without constraints, that the transition of frontiers is easy, or that the operations of the human intellect are arbitrary. As in Bruno's cosmology, every single world in the universe is finite, and most events happen in this finite frame. These finite, local frames are infinitely embedded in larger frames, and humans can conceive no absolute frame.

Bruno's assumption that all other solar systems were identical to the system where we live was too optimistic. Our solar system at the rim of our galaxy is unique, and the constellation of the moon and earth as a tandem may be exceptional. Nevertheless, if the universe is infinite, there should be other solar systems similar to ours, and it is highly probable that they have developed or will develop intelligent beings. In this very general sense, symbolic forms have a cosmological frame that surfaces in the symbolic form of myth and religion discussed in Chap. 5. The most pertinent local frame is the ambient ecology of humans. In many animals, this frame is severely restricted. Evolutionary changes in the climate and the surrounding ecology can radically change this frame. Human evolution was and still is dependent on such changes. The human species can even transform the frame of its survival through its growth and the industrial exploitation of the ambient ecology. The concept of "Anthropocene" points to this aspect.

The human mind and its media of communication can transgress given frames. In a larger sense, this was also true for humankind on its path from biological to cultural evolution, a story of frame changes and extensions. Therefore, different (genetically or culturally given) frames constitute human consciousness. In the following section, three outstanding frames (called "ontologies") are distinguished: (1) A cosmological frame that, for practical purposes of human communication, can be focused on our world, the earth, and the spatiotemporal surroundings of humans. (2) The human body (the mind and all body parts) is relevant for human communication (active and receptive). (3) Insofar as humans are "social animals" (cf. Aristotle's "zoon politicon"; class. greek: ζῷον πολιτικόν), all social entities beginning with the mother–child-group are relevant frames of human communication and symbolic forms. As in other classification cases, we suppose an underlying continuum, and thus an infinity of possible frames. It is a question of parsimony to assume a finite and small number. The notions of sign in antiquity, in Augustine, the middle ages, and most modern theories of the sign (e.g., the tripartite distinction by Peirce) and the shortcomings of reductionist theories that eliminate either the first frame (the world and the reference to it) or the social dimension (e.g., cognitivist models) or reduce the body (frame 2) to the brain are arguments for such a tripartite ontological framework.[2]

---

[2] The term ontology does not presuppose a metaphysical assumption about the universe's structure. Every ontological frame is a practical postulate that helps to organize our understanding of basic processes in perception, cognition, knowledge, and social action. In this chapter, the term "semiotics" includes the domain called "philosophy of symbolic forms" by Cassirer [5]. Symbolic forms

7.2 Science in the Spirit of Kepler and Peirce and Semiotics

In his "Lessons on the History of Science", given in the context of Lowell Lectures in 1892, Peirce characterized Kepler's intellectual strategy:

> His admirable method of thinking consisted in forming in his mind a diagrammatic or outline representation of the entangled state of things before him, omitting all that was accidental, observing suggestive relations between the parts of his diagram, performing diverse experiments upon it, or upon the natural objects and noting the results.
>
> The first quality required for this process, the first element of high reasoning power, is imagination; and Kepler's fecund imagination strikes every reader. But "imagination" is an ocean-wide term, almost meaningless; so many and so diverse are its species. What kind of imagination is required to form a mental diagram of a complicated state of affairs? Not that poet-imagination that "bodies forth the forms of things unknown",[3] but a devil's imagination quick to take Dame Nature's hints. The poet-imagination riots in ornaments and accessories, a Kepler's makes the clothing and the flesh drop off, and the apparition of the naked skeleton of truth to stand revealed before him (Peirce [21]: 255).

In his analysis of the trajectory of the planet mars around the sun, Kepler had to imagine possible mathematical laws that correspond to the data Tycho Brake had collected until he died (cf. [14]). Currently, mathematicians and physicists develop devices called "The machine mathematician" that rehearse the work done by Kepler, i.e., they extract equations from a large set of data. The first algorithm, called BACON, was able to "find" the Kepler laws in astronomy. An advanced algorithm is called "Symbolic Regression". Kepler's "imagination" is simulated by considering a large set of possible mathematical formulas, searching in different directions, and using a kind of Darwinian competition between different routes.[4] Applied to semiotics, the tradition of Eco, Greimas, and others restricts a priori the choice of mathematical tools to simple logical patterns (cf. the dominance of binary oppositions and the universal use of the "semiotic square" by Greimas and his followers). In practice, they adhere rigidly to these simple devices and arrange or interpret the data available such that a plausible account may be reached.[5]

Peirce's observations on Kepler foreshadow the history of semiotics in the twentieth century. Some of Saussure's followers in France in the sixties developed a literary style of structural semiotics, which was quickly absorbed into studies of mass media, journalism, and modern media sciences. Umberto Eco (1932–2016) was the

---

are signs in the sense of Peirce. In addition to Peirce, the arts, music, myth, and religion are dealt with on par with language and scientific communication. Semiotics is more than a generalized logic or linguistics.

[3] A quote from Shakespeare: "And as imagination bodies forth//The forms of things unknown, the poet's pen // Turns them to shapes and gives to airy nothing // A local habitation and a name". William Shakespeare, A Midsummer Night's Dream.

[4] Cf. recent advances in "machine imagination". Thus Li, Ji, and Zhang [16] use Artificial Intelligence to describe the scientific itinerary from Kepler to Newton.

[5] The source of this attitude was probably the success of Chomsky [6] and his restriction of the rules of grammar to set theoretical notions and the hierarchy of rewrite rules in the so-called Chomsky hierarchy.

most prominent author in this field. He was the director of the Bologna center of mass media and a successful author of novels. He stood firmly on the ground of a philosophical-literary culture in Romance countries but sought theoretical backing in the Danish school of structuralisms (mainly Hjelmslev). In the last decades of the twentieth century, Eco took notice of American cognitivism, mainly cognitive semantics. In these developments, semiotics became a Janus-headed discipline. In its practices and mainly in its teaching in Europe and Latin America, semiotics in the spirit of structuralism (Eco, Greimas, and followers) stuck to the literary and philological traditions and avoided the difficulties of formal reasoning and strict methodologies in empirical research. In theory, it had the ambition to cope with modern interdisciplinary science but could not go beyond theoretical eclecticism. Coming back to Peirce and Kepler, this major trend of semiotics opted for a poet's style of "scientific" imagination and against the "devil's imagination quick to take Dame Nature's hints" (cf. the quotation from Peirce above). Semiotics should instead pursue the "devil's method" using advanced conceptual tools in modern mathematics, e.g., dynamical system theory. Insofar as these tools have been developed and applied in the natural sciences (physics, chemistry, biology, cognitive sciences), semiotics in this style can be called "semiophysics" (cf Thom [25]). In the following, the focus will be on the epistemology of such an enterprise.

In the humanities and specifically semiotics, this endeavor should reflect on a proper ontology, i.e., the fundamental assumptions or foundational concepts supporting the theoretical construction. One needs three ontological premises: the world we experience (W), the individual sensitive and cognitive system (I), and a community/culture/society (C) as the place of coordination,[6] shared knowledge, and tradition.

7.3 The Threefold Ontology of the World (W), the Individual Mind (I), and the Community of Humans (C)

A central question in semiotic studies is: Can humans know the world, themselves (their minds, indirectly other humans and their minds), or at least have an approximate idea about these matters? Two basic approaches to knowledge can be distinguished: via the world or self-experience (reflection). Both approaches must be brought to a synthesis in the individual mind and shared with a community, which is the locus of selection, stabilization, and objectification of individual insights, and thus the decisive factor in gaining relevant and persistent knowledge. The relationship between

[6] The phenomena of coordination in inanimate and animate dynamic systems are the central topic of "Synergetics", developed by Hermann Haken, Scott Kelso, and others, cf. [10] and [15]. These models and mathematical treatments are always the background when in the following chapters, coordination is put to the fore.

the individual mind and the collective "mind" is the bottleneck of objective knowledge. Many proposals for the solution to this problem have been made in the history of epistemology.

A one-sided emphasis on the collective aspects considers knowledge exclusively from the perspective of academic institutions (schools, universities, research institutions, and accepted traditions).[7] It reduces the problem to the formal and political organization of the communities, ignoring individuals' experience, curiosity, and intuition.[8] Individual thinkers do not only commit themselves to an intellectual direction; they contribute to it, carry it and fill it with life, starting from their personal experience. Intuition, experience, and reflection, however, seem to be not communicable because they presuppose some comparative instance (a "tertium comparationis") that can catalyze personal experiences. Nevertheless, similar individual concepts can be matched using identity-preserving transformations (automorphism).[9] The individual thinkers must compare their experience and reflective insight with the experience of others (i.e., they should achieve a third-person perspective) and eventually transform it into a general system of signs, such as linguistic, visual, technical, or other signs. On this collective level, the others can then understand the personal insight of an individual.

All three entities, the world (W)–the individual (I)–the community (C), are in motion, changing continuously. The individual (I) experiences and constructs the world (W) in thought. The world, including that of others, who belong to the self's environment, is resistant, i.e., the construction of the world by the individual is evaluated, shaped, and reshaped through interaction with the world and others in everyday actions. While the individual perceives the others who also interact with the world, he participates in their world construction, at least under essential aspects, and coordinates his world perception with that of the others. This selective participation only affects those aspects relevant to the communal experience. There are, for this reason, many individually different world constructs, which, however, are "heat-sealed" at the seams of collective interest, at the points of coordination. The constellation of three correlated ontologies forms a very complex system. In mathematical terms, specially developed for astrophysics, it is comparable to a three-body

---

[7] Non-human collectivities, i.e., cooperating social animals and integrated assemblies of plants or other biotic entities, show similar features. The morpho-dynamic models of Thom/Zeeman and the synergetic models of Haken/Kelso have the advantage that the concepts they apply do not presuppose human conscience and free will.

[8] In the case of individual animals or even plants, similar arguments can be formulated. Thus, Darwinian selection operates basically on the level of individuals who survive or procreate.

[9] Automorphism is "mapping the object to itself while preserving all of its structure" (Wikipedia). In color perception and color labeling by individuals, a color space is defined with a specific topology and relations of isotopy. Kleiner [13]: 20 writes: "Putting everything together, we conclude that any statement, scientific or otherwise, that addresses color experiences sensu stricto—i.e. which addresses what it is like to experience colors—only makes sense if it is invariant with respect to Iso(E) [equivalence preserving; W.W.] transformations. This is a consequence of the fact that qualia are non-collatable and of the corresponding freedom of every experiencing subject to choose names for the qualia he/she experiences."

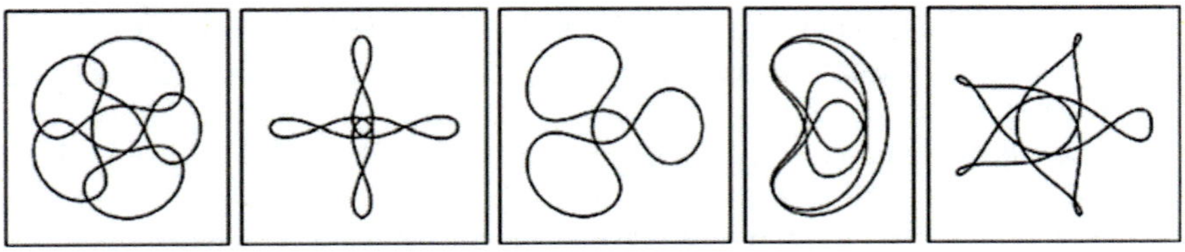

Fig. 7.1 Orbits of a planet in a binary star system

problem. How can the three movements, the world, the individuals, and the community, be combined to form a stable or at least metastable overall system? Poincaré has already shown that the three-body problem can only be solved in exceptional cases. In general, it shows chaotic dynamics (see Montgomery [19]). We cannot elaborate on this analogy and possible mathematical models in this context. After all, it shows that a seemingly simple configuration of three force fields leads to an almost uncontrollable dynamic. The two-body problem in one dimension can have the character of a linear interaction between two billiard balls that collide. In two dimensions, either two bodies of similar size orbit a center (e.g., in a binary star system), or a smaller body (a planet) orbits the larger one (e.g., the sun). Li and Liao [18] simulated the problem and possible solutions,they found 695 examples of coordinated movement of three bodies with equal mass. In any model of this interaction, it is crucial either to reduce the three-body problem in a rough approximation to a two-body problem or to consider particular constellations in which the movement pattern of the three bodies (centers of force) is periodic or almost periodic and thus describable (understandable).[10] Figure 7.1 shows "some of the possible repetitive orbits of an idealized planet moving in the plane of a pair of stars that are in a perfect elliptical orbit" (Wolfram notices, page 972: https://www.wolframscience.com/nks/notes-7-4-three-body-problem/.

A minimal scenario of insight and processing in cognition (perceiving, feeling, categorizing, or evaluating) has the following components:

1. A human environment is recognizable by the available sensory apparatus and cognitive centers. We call it the world (W). Under the condition of sensory and cognitive accessibility, this world is different for each living species and varies (less dramatically) between the individuals of a species. Thus, there is also a reality beyond this world accessible to our species. Man has succeeded in expanding the area of his insight and knowledge again and again; nevertheless, the world (W) remains fundamentally different from reality (the thing-in-itself, according to Kant). The human environment naturally includes living beings, especially other people. With these, mainly if they use the same system of signs, the individuals can exchange ideas about their world knowledge. Knowledge of

[10] The dyadic conception of signs by Ferdinand de Saussure reduces the traditional sign triad (St. Augustine) to the duality of"significant = image acoustique" versus "signifié = image mentale". Thus, it simplifies the three-forces-constellation to a two-forces-constellation which is not in the neighborhood of chaos. Concerning state-of-the-art in 1900, this was a promising scientific strategy.

the world is, thus, to a certain extent (not generally), dependent on the shared system of signs. This part of the world (W) forms the reference for the use of signs and, thus, the necessary point of departure for every semiosis.

2. Although human individuals and sign users have genus-specific biological properties, they are individually different. Moreover, their "world" is developed differently due to specific living and learning conditions. Nevertheless, individuals are the actual place of perception, cognition, and communication, so we must use them as fundamental centers of force in our dynamic modeling. We symbolize this center with the individual (I); the number of individuals can be huge (up to the number of people living in this world). These individual worldviews can be the reference domain of signs in communication; i.e., individual worldviews or perspectives can be and often are the proper reference basis for signs. In many languages, the traces of this ontology are constructions containing "propositional attitudes" like "I think, I believe, I doubt that".[11]
3. Communication and the use of signs always involve several individuals, i.e., a model that only has the individual and the world as a reference is fundamentally inadequate. The network of individuals has a rich structure because there are many individuals and many kinds of intersubjective or communicative relationships. From the point of view of the individual, these networks also appear differently. Besides, there are global structures in society that are no longer transparent from the individual's perspective and thus only rarely influence his behavior or do this only in extreme situations. The social environment ("society") is a sine qua non of all communication and the use of symbols. In this respect, although it appears abstract and opaque from the individual's perspective, it must be recognized as a primary force, i.e., as an anchor point of the semiotic model. In particular, systems of behavioral norms, conventions for the use of signs, traditions, and their validity beyond the lifespan of individuals cannot be explained without accepting this ontology. This necessity is particularly true of symbolic forms (in the sense of Cassirer). We symbolize this component as a society or community (C).

As the above comments have already shown, the three forces depend on one another; one can call the dependencies force fields and consider the topology and dynamics of such a field. The force fields that connect them are to be interpreted in the simplest case as an attraction (positive) or repulsion (negative). This analysis results in a plausible analogy to physical facts and furnishes the motivation to use the corresponding conceptual techniques in semiotics.[12]

In the triple constellation presented, the world is by far the most potent component in that it enables both the evolution and the continued existence of individuals

---

[11] In the case of language, this is evident for speech acts, modalities, and propositional attitudes. In visual art and music, neglect of this dimension would make these symbolic forms trivial and lacking in content.

[12] This decision is in no way a physicalist reduction because the conceptual tools used in the natural sciences are symbolic forms specifically tailored in the context of physics, chemistry, and the life sciences.

and the constitution of society. Compared to the changes in the individual (I) and society (C), the world (W) is relatively constant, i.e., it changes more slowly, and the influence of (I) and (C) on (W) is relatively small. This is because individuals and societies are parts of the world, just as individuals are parts of society. However, there is another hierarchy relative to (I) and (C). The individuals (I) are biological-evolutionary entities, and the societies (C) are either organized in the individuals' mental representations or indirectly in the artifacts such as settlements, traffic routes, architecture, laws, norms, etc. Therefore, they are dependent, secondary entities that have been consolidated or gained power in the course of history.

In the context of the analogy mentioned above between the astrophysical system and the communication system, the world (W) plays the role of the sun, and the individuals (I) are satellites of the sun like the earth or other planets. The societies are like a planet's moons (cf. Fig. 7.2). This planetary image is, however, only a mnemonic crib that may help to illustrate the network of dynamic relations between the three frames and to serve as the base for the use of a semiotic concept of force, force field in this network. Further research must show how far this metaphor can lead us. Less metaphorically, we can state: (W) is the base of all sign objects, (I) is the realm of secondary, and (C) is the realm of tertiary sign objects.

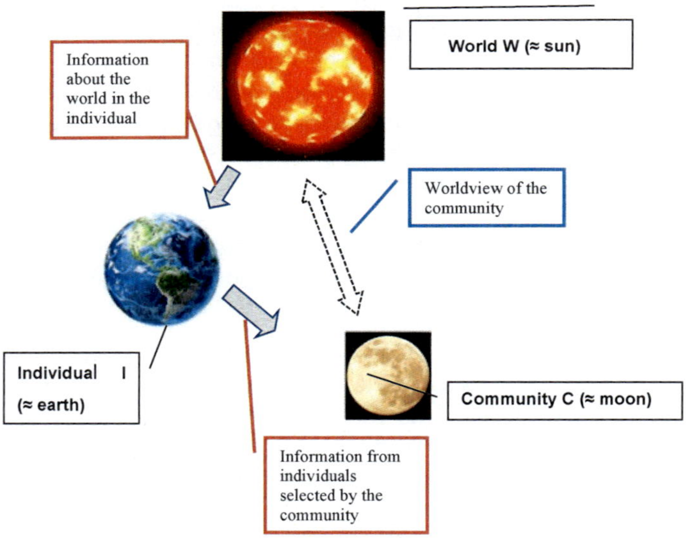

Fig. 7.2 The constellation world—individual—society and information transfer in a three-body system

7.4 Cycles and Hypercycles in Semiogenesis

Based on the Kepler ontology sketched earlier, we can use a metaphorical link between the realm of astronomy (Kepler's primary area of intellectual engagement) and the threefold ontology of semiosis. This metaphor may seem farfetched, but it allows for a simplification and systematic conception of the rich relational networks inside the threefold universe of symbolic forms (signs in terms of Peirce).

The basic metaphorical equation is planetary center/orbits $\approx$ domains of semiosis. On the side of the planetary system, three agents (force fields) are distinguished: The *sun* has a gravitational force that governs the orbit of the planets; it illuminates them and provides them with energy (of radiation). The *earth* is in motion around the sun and follows specific laws of motion (Kepler's laws); it holds the *moon* in its orbit and is illuminated by the moon at night. Finally, the moon has a fixed orbit and governs the dynamics of the tides (and other processes) on the earth. In the semiotic and epistemic universe, we can distinguish similar cyclic processes:

1. The evolutionary (Darwinian) cycle governs the evolution of life and the tree of species. Singularities like the impact of planetary bodies on the earth, physical catastrophes like volcano eruptions, and climatic crises play a prominent role. Our species, the Homo sapiens, is a late product of this evolutionary cycle. The driving force is the confrontation with the world (W), i.e., selection on the background of mutations (genetic "noise").
2. The individual mind is, together with the whole body, a product of the first cycle. In the biography, learning, and other adaptation cycles, the individual (I) achieves a level of knowledge and conscience, including symbolic forms. This development is cyclic insofar as significant experiences are repeated over and over again and create a system of knowledge, which serves as a stable background for cognitive patterns (cf. Chap. 6). The driving forces are experience and knowledge. They are constitutive of the individual mind (M).
3. The collective "mind", i.e., the selection and accumulation of the experiences and memories of many individuals in a community, is insofar cyclic as each social contact; each exchange of experience or memories and knowledge contributes to a cycle of selection (reduction) and synthesis (augmentation) in an infinite process. These processes constitute the community/culture/society (C).

Together, the three cycles form a hypercycle in the sense of Eigen and Schuster [8]. They have elaborated this notion in the context of a model that should explain the origin of life. In Wildgen [31], it was adapted to explain the origin of language. Hypercycles are rapid morphogenetic processes in biochemical, biological, and symbolic systems. In Sect. 6.1.3, the hypercycle concept has been applied to understand the rapid evolution of human language capacity.

7.5 The Dynamics of Information in the Threefold Ontology: World—Individual—Community

The informational attraction between the three centers (W), (I), and (C) is based on the interplay of *salience* and *pregnancy* fundamental for the human cognitive apparatus (cf. Thom [25]; Wildgen [32], and Sect. 1.4 in this book). Salience concerns the perceptual apparatus's selection of external impressions (stimuli). They are either ignored if the perception window does not capture them, accepted if the perception threshold is crossed, or reinforced if the stimulus optimally fits the physiological filter. Vital relevance or "pregnancy" ("prégnance" in Thom [25]) has its source in the vital needs of the individual. Perceived impressions are rated either as particularly relevant or as meaningless. As the product of these two instances, a specific effect emerges, leading to action or enabling a stronger or weaker memory entry. The overall effect can be called semantic information.[13] The individuals (I) obtain information from the world (W) and exchange it selectively with one another, with the help of a medium, e.g., an acoustic, a visual, or another symbolic system. A selection can inform the community (C), which stands, therefore, indirectly in a relationship of information to the world.

The informational attraction of individuals by the world is the strongest force because incorrect perceptions and evaluations negatively affect individuals. In contrast, successful information transfers are associated with positive effects. Overall, a chain of reinforcements and hindrances emerges and forms a profile of the world for the individuals (compare the concept of "affordance" in ecological psychology; e.g., Gibson [9]). Every exchange between individuals already constitutes a social reference, i.e., the starting point of society. Society emerges in a synergetic process in which individuals are differently involved. The process is very selective, i.e., only highly rated facts about the world are used in forming opinions and decision-making in society. Indirectly (see the perforated arrow), a phenomenon arises that Wilhelm von Humboldt (1767–1835) called a "worldview".

In the case of myth/religion, global ideology, and general scientific models (e.g., the big bang model of astrophysics), the following features are noticeable and differ from everyday communication:

(a) The salience of features that go beyond, i.e., "transcend" everyday experience (e.g., theoretical entities in physics, ethical or religious principles, social utopias) can be minimal or even zero since there is no confirmation of these transcendent entities in sensory experience. The pregnancy, i.e., the expected value for knowledge or human consciousness, is, in many cases, high. Even (imagined) traces of a sensory salience can suffice to form stable signs and sign systems.

[13] Cf. Carnap and Bar-Hillel [3] for the notion of "semantic information" and its application to semantics (and sociolinguistics) in Wildgen [26].

In this context, information can emerge from (almost) nothing, i.e., imagined facts are assigned a quasi-sensory salience.[14]

(b) Selection and reinforcement in the transition from individuals to society are decisive for accepting general principles in science, religion, or ethics. The social coordination of opinions and evaluations transforms weak individual causal attributions into firm convictions and normative articles of faith. At the same time, these constructs are the skeleton for stable social rules and norms.[15]

(c) The indirect relation to the world in the case of societies can lead, among other things, to the fact that some properties of the directly perceptible and assessable world that are relevant for survival or well-being are ignored. A society can existentially fail, caused by its generated ideological or religious phantasms. It can destroy itself or lose strength and the courage to survive in competition with other societies. However, the opposite can also occur, for example, if the state of the world is highly discouraging and resignation and fatalism spread. As a remedy, an optimistic, religiously, or ideologically motivated, although unrealistic, view of the world can help through the time of need and avert a downfall. Cf. for the dynamics of myth and religion Chap. 5 and Wildgen ([36], in German).

The fundamental instability of the three-body constellation has immediate consequences for the stability of symbolic forms. Specific solutions, i.e., at least temporarily stable scenarios, will either reduce the number of relevant force-centers and result in a stable two-body constellation or reduce the information exchange to a partial and limited exchange between two or three of the bodies. The second solution corresponds in the astronomical context to the rigid relation between the moon and the earth, insofar as the change and development of a society are restricted in comparison with individuals, similarly as the moon's orbit is restricted insofar as it always shows the same side to the earth.[16]

In the case of religions, ideologies, and other global systems of belief, three scenarios may be considered (along with many others imaginable):

- Every individual tries to make sense of the world he/she perceives, experiences, and in which he/she acts. Shared knowledge and faith tend to disappear under the condition of individual fragmentation. This situation corresponds to an "anomic community" in the sense of Durkheim [7]. If one looks at the distribution of knowledge and opinions in communities, one can observe a natural tendency

---

[14] In mystical empathy, sensations can be generated that are very intense. In some cases, they can even have dramatic physiological consequences, such as the opening of wounds on the hands and feet in the imagination of the sufferings of Christ.

[15] The religious "information" is objectified in material forms, which can become the object of salient and pregnant perceptions and thus create a second-order relationship to the world. In addition, through artifacts such as images, sculptures, religious rituals, or texts, religious belief can be experienced with the senses. However, these secondary realities may lead to the prohibition of sculptures, images, and names (cf. Wildgen [36], Part II).

[16] Galilei showed that the earth and other planets (in his case Jupiter) have a moon or several moons. Meanwhile, astronomers found many planets and their moons outside our solar system.

toward individualized and context-relative knowledge and belief. The dynamics of individual variation are never absent but just concealed by the prominence of institutionalized norms.

- The impact caused by the individual is reduced or eliminated. Totalitarian ideologies and societies tried to achieve this goal. In periods of conflict, war, or global repression, such a "solution" may be preserved even for some decades. The Spanish Inquisition (main period 1500–1700) and the quasi-monarchism of the pope in the context of the 1870 dogma of infallibility tried to focus religious authority on one institution or its highest representative. More recent examples are Stalin's Russia between 1927 and 1953 and Hitler's Germany between 1933 and 1945. Modern neo-Islamic "Jihad", i.e., a war against all non-Islamic religions and populations, tries to eliminate any information exchange between the different religions. It hinders the information exchange with individuals that do not belong to the same community. As intellectual and also commercial information exchange is the motor of any successful governance in the globalized world, such a strategy cannot contribute to the welfare of the members of these societies, and as soon as war-like situations disappear and mass murder loses its acceptability, such "solutions" will also lose their attractiveness.
- The most radical reduction annihilates the role of the world; insofar as it does not fit a given grid, a fixed worldview. In the modern technological world, this attitude is only possible if the area of technology and underlying natural sciences is exempted from such blindness. Modern warfare depends essentially on technology, which leads to a strict separation of technical and military information from social or political information. The first is considered highly relevant to society and its leaders. All kinds of knowledge, which on one side are not relevant for warfare or even dangerous for the "frozen" ideology, are repressed or frozen in an earlier standard.[17] Philosophy, ethics, sociology, psychology, and communication sciences are reduced to interpreting religious or ideological texts. In the context of Christian denominations, the Copernican hypothesis, the philosophy of "free" thinkers in the seventeenth and eighteenth centuries, and the Darwinian hypotheses were excluded from any information transfer in specific communities. As interdictions had no immediate impact on current economic and social life, they could be sustained for long periods. Eventually, they endanger the coherence of general education and the growth of scientific knowledge and produce a kind of "backwardness" with negative economic and political consequences.

[17] George Orwell describes the role of knowledge versus ignorance in societies preceding the utopian society described in his book "1984" as follows: "In the past, also, war was one of the main instruments by which human societies were kept in touch with physical reality. All rulers in all ages have tried to impose a false view of the world upon their followers, but they could not afford to encourage any illusion that tended to impair military efficiency. [...] In philosophy or religion, or ethics, or politics, two and two might make five, but when one was designing a gun or an airplane they had to make four. [...] War was a sure safeguard of sanity, and so far as the ruling classes were concerned it was probably the most important of all safeguards." (Orwell [20]: 206). In Orwell's dystopia, even wars are arbitrary fictions, such that the poor remains of sanity are abolished.

7.6 The Information Flow Between the Forces: World, Individual, and Community

In Fig. 7.2, the arrows joining the three bodies were titled "information" and "informational attraction". The notion of "semantic information" of Carnap and Bar Hillel [3] supposes a world represented by logical propositions. In such a logical space of possible states of affairs, information is the measure of restriction to subsets of states. The information increases with the amount of restriction. It is zero (0) if all possible states remain possible (are not reduced) and (1) if only one state is left, i.e., if the state of the universe is finally specified. This concept of information is rudimentary and abstract. Therefore, it is clear that the three bodies considered, the world, the individuals, and the society, ask for different information codes in the constitution of a space representing their state of affairs. In the following, (I) shall sketch the nature of such universes of information and their interaction:

- A society with its norms, rules, commands, and interdictions easily fits into the propositional logic framework. For illustration, we may consider the list of commands and interdictions in Judaic rules of Shabbat and food prescriptions. There exists a finite list of propositions describing the content of commands (you should follow p) and interdictions (negative command: you should not follow p). If all commands and interdictions are followed, the believers are in perfect concert with their society (their religious community) and obtain the maximal information value (1). On the other hand, if they break all interdictions and do not follow any of the commands, they reach the information value of zero. This propositional model of the universe of information can be adapted to any religion or law-governed social system. Suppose this type of information is projected on the world (W). In that case, one can imagine a world of perfect obedience to the religious (social) rules, i.e., heaven, and a world of total disobedience, i.e., hell. The quasi-logical structure of these religious concepts is probably not natural or primal for religious thought but rather a product of logo-centric reconstructions following the elaboration of rhetorical and argumentative schemes developed in literate antique societies, i.e., after 3000 BC[18] Cf. Chap. 5 and Wildgen [36]: 140–179) for more details.
- Information flows in the individual mind were recently (mainly in the second half of the twentieth century) approximated by the consideration of "logical" machines, i.e., computers, that, on their part, are based on logical calculus. Since the end of the twentieth century, it became evident that the "machine in the wet", i.e., the brain, follows different principles. PDP models (PDP = parallel distributed processes), neural networks, neuro-computers, and integrated information theory (IIT) looked for alternative concepts. The fact that language plays such a dominant role in human thinking suggests that specific (logo-centric)

[18] The oldest cuneiform and hieroglyphic scripts are dated to 3400 and 3200 B.C. However, their religious impact may have occurred later in the context of mythical narratives' fixation and ritual instauration.

aspects of human thinking can be translated into a propositional format. But what about human neural procedures of imagination and thinking, spatial-dynamical, and continuous-fluid aspects (cf. emotion and evaluation)? Are they lost in the information exchange between individuals and their society? If this is the case, do individuals accept this dramatic loss? A significant problem exists already at the micro-social level. If two persons, e.g., a couple, exchange information, are the spatial-dynamical and the continuous-fluid aspects of information lost? Probably not. However, if this is so, one can conclude that, at least in perspective, these aspects can also be transferred first to the family, the group of friends, the networks of regular social contacts, etc. With social distance, the loss of a shared spatial–temporal reference base, and the insecurity of transmissions of analog contents, the information flow may be biased toward the more rigid, two-valued propositional schemes. Consequently, such global communities lose their human relevance or even become anti-human. We can conclude that the concept of semantic information based on propositional logic is at least partially inadequate and must be enriched by non-propositional concepts of information and information transfer based on continuous scales (analogous information) and spatiotemporal information (theoretically assessed in dynamic systems theory).

- The problem of information flow becomes even more dramatic if we consider the flow between the world as we perceive and experience it and the individuals. The history of the natural sciences illustrates this problem alongside possible routes to its solution. Galilei said that scientists should read the book of nature and that this book is written in terms of mathematics. This persuasion is based on the fact that modern natural science was able to advance helped by specific languages tuned to exact observation and strict inferences, i.e., mathematics, and that in this context, the sciences were able to develop technics of measurement, instruments of observation (e.g., the telescope and the microscope) and strategies able to control the use of masses of observations and measurements by many scientists over the world and to communicate/discuss them effectively. "The book of nature is written in mathematical terms" was a dramatic simplification by Galilei. Still, the scientific community has found an effective way of cumulating individual experience into a flow of information that gives access to knowledge about the world and has allowed controlling it to certain limits. The scientific information flow can be transferred to the overall population (with losses and retardation) and via the transfer mentioned above between individuals and society to the shared knowledge of a society. The conflict between science and religions that represent more or less accurately a state of knowledge some thousand years ago (in the time of Abraham, Jesus, or Mohammed) and modern science since Copernicus reflects the difficulties of a coherent and continuous information transfer between the world (W) (experienced and scientifically understood), the mass of human populations, and the social or religious entities, which make a selection out of these inputs.

The critical phases of religions, ideologies, and to a certain extent, scientific theories concern the transition areas in these information flows and the rigidity of many knowledge systems. A solution to this problem asks for semiotic management that is aware of the dynamics inherent in the three-body constellation of (W) (world)–(I) (individual)–(C) (community, society).

If we have added the third ontological frame (C) (community, society) to the traditional dual constitution in Cartesian epistemology, this was also a consequence of Cassirer's reorientation toward culture, a move that was biographically linked to his encounter with the Warburg circle in Hamburg. However, already Cassirer's historical analyses of Giordano Bruno (1548–1600), Giambattista Vico (1668–1844), and Wilhelm von Humboldt (1767–1835) in Cassirer (1922) show that the tradition of a philosophy of culture was in the center of his interests. We must, therefore, reflect on the genesis of the Philosophy of symbolic forms out of this tradition of cultural studies.

7.7 Cassirer's "Philosophy of Symbolic Forms" in Its Historical Context

Cassirer's philosophy aims at a philosophy of human culture that addresses symbolic forms such as language, myth, science (later art, ethics, and technology). The meaning of the symbolic forms is given by the internal "reference structures" of consciousness; i.e., mental entities of different provenience and character are brought together. This trend is fully developed in human cultures and concentrated in a plurality of symbolic forms. The analysis of this plurality and the interrelations between the basic types is at the heart of Cassirer's theory of culture.

The position of language between scientific forms of knowledge and other forms of culture, such as myth and art, can be traced back to the fundamental opposition between representation and intuition. The representation most clearly expressed in scientific theories (and their forms of language) risks losing the immediate reference to human life because of its abstractness and complexity. A scientific language's indirectness and intellectual instrumentality create an increasing distance to the archetypal human intellect (Kant), and the symbolic form runs the risk of becoming blind thinking (Leibniz). Pure intuition, as sought by mysticism, reduces the distance by eliminating the mediating, higher organized symbolic forms and risks becoming "speechless", i.e., it leads to illusionary knowledge and a kind of solipsistic "communication".

In the third volume of the Philosophy of Symbolic Forms with the title:: "Phenomenology of Knowledge" (Cassirer [5], vol. 3), some topics from the first and second volumes are taken up again and updated. In particular, the philosophical contributions of Husserl, Bühler, and Heidegger are incorporated. Weyl (group theory), Hilbert (axiomatization), and the Erlangen program are considered in questions of mathematical form. Several specifications of Cassirer's philosophy of symbolic forms can be discerned:

- Mathematics and the mathematized natural sciences allow for the analysis of "pure meaning". Despite Cassirer's objective that the philosophy of symbolic forms should be a philosophy of culture and his skepticism about a rapid transfer of results in physics to psychology and other humanities, Cassirer never gave up the idea of the unity of (scientific) knowledge and the guiding function of mathematics.
- The basic ideas of Husserl in his books: "Logical Investigations" [13] and "Ideas for a pure phenomenology" [12] are condensed by Cassirer (ibid. 228–232) in the formula: to have sense = consciousness "because "to have sense" or "having in mind" is the basic character of all consciousness, which is therefore not just an experience in general, but a meaningful, "noetic" experience" (ibid. 230). Husserl is criticized because (in Cassirer's analysis) he reintroduces the Cartesian dualism and separates form and content within consciousness.
- Cassirer comments on Heidegger's book "Sein und Zeit" (Heidegger [11]) in a lengthy footnote (Cassirer [5]: 173 f.). Heidegger seeks a foundation of human thinking in those entities that are at hand, "das zur Hand Seiende" (the "handy, at hand"). However, human thought can only reach the "Sein" (existing) via the path of symbol formation, i.e., symbolic forms are the turning point of human knowledge.
- Cassirer accuses Kant of expressing his transcendental insights with concepts of eighteenth-century psychology (ibid. 226). In Kant's psychological analysis, the "mind" acts like a sorcerer or necromancer who animates the "dead" sensation "and awakens it to the life of consciousness" (cf. ibid. 227).

According to Cassirer, human intellectual history unfolds different aspects of language that, depending on the preferred perspective, are identified with science/mathematics, music, or art. These tendencies in reflection on the mind and culture suggest a close relationship between the symbolic forms mentioned. Therefore, a philosophy of symbolic forms should reveal common principles. The specific theories of symbolic forms, language, myth, science (and art, ethics, and technology) should reflect the framework of a philosophy of symbolic forms, describe the differences and transitions and capture the achievements and functions of the types of symbolic forms.

7.8 Philosophy of Symbolic Forms (Cassirer) or Generalized Logics of Signs (Peirce): A Confrontation

Cassirer reoriented himself toward a philosophy of culture during and after the First World War (perhaps in connection with his appointment to the University of Hamburg and his contacts with the Warburg circle). Peirce died in 1914, but decisive writings and lectures originated before and after 1900. Peirce and Cassirer are excellent connoisseurs of intellectual and scientific history since antiquity. For both, German

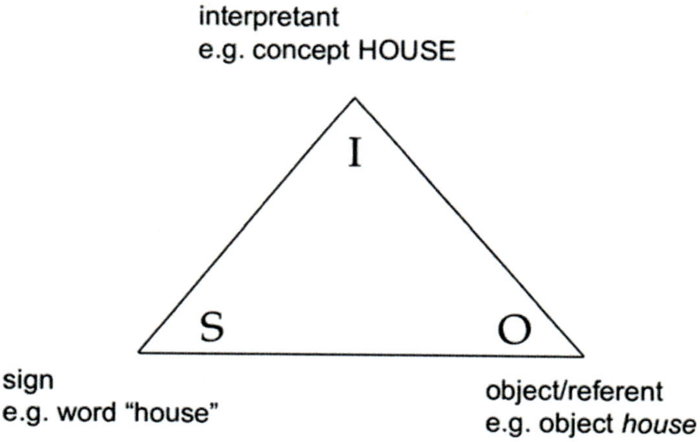

Fig. 7.3 The triangle of sign functions with examples from language use (English)

idealism, especially Kant, is a point of orientation. After 1920, Cassirer aimed at a philosophy of culture and brought the evolutionarily and historically fundamental forms of myth and language into competition with the formal disciplines of logic and mathematics.

Consequently, he abandoned the strictly ordered architecture of ideas and allowed for parallel and independent cultural forms, thus the plurality and not hierarchically ordered set of symbolic forms. Peirce had also considered a plurality of different sign systems and their interaction (Cassirer became only aware of Peirce's opus in the forties when the Collected Works of Peirce began to be published; the first volume was published in 1931; cf. Peirce [22]).

The triad of sign functions, according to Peirce, contains *sign form* (representamen; S)–*sign object* (designated; O)–*interpretant* (thought, content, inner image; I) (Fig. 7.3).

The notion of interpretant in Peirce's writings has many aspects. The *immediate* interpretant is the idea of the object in the mind of the sign user. Peirce calls it *degenerate* because it neglects communication with other users and its effect on them.[19] The *dynamic* interpretant focuses on the effect of a sign in another mind and the real consequences of these forces, including contingent factors. The notion of interpretant (I) is on another level of complexity than the duality of the object (O) and the sign (S). In the notion of interpretant, the mental meaning or image in the individual is completed by the other mind. It has a dialogic and pragmatic dimension. Compared to the three-folded ontology sketched in this chapter, the sign concept of Peirce takes into account all three levels.[20] However, Peirce's triangular constellation

[19] In solitary thinking or inner monologue, a communicative act is only simulated; but it presupposes the normal usage of signs in social exchange.

[20] Saussure only distinguishes the signifying (French: "signifiant", the mental image of the sign, e.g., the word, picture, musical theme, etc.) and the signified (French: "signifié"; the mental image

does not mirror their interaction's complexity. The notion of "interpretant" only "labels" the dynamics which manage this relation. Still, the forces that govern it and the creativity of semiosis in the interaction between S, O, and I are neither specified nor a theoretical account given.

Ernst Cassirer has tried to attack this problem in his "Philosophy of Symbolic Forms" [5].[21] Symbolic forms are based on "conceptual knowledge" and "experiential knowledge". Both levels can be integrated, leading to a complex synthesis demanding human creativity. At the experiential level, the flow of perceptions and reactions is canalized into pregnant forms, i.e., individual experience is shaped by limitations, transitions, and values of relevance; it is the result of semiotic morphogenesis at the individual level. At the conceptual level, the dynamics involve more than individual mental processes; the relevance profiles of creative semiosis involve socio-semiotic and historical morphogenesis and self-organization.[22] Both levels neither describe (mirror) nor construct reality; "Cassirer understands the experience and knowledge of reality […] as an intelligent orientation in *reality*" (Plümacher [23]: 169). This orientation depends on contexts, perspectives, and values. As a consequence of these insights, we distinguish two levels in the schematization of semiosis: an epistemic level (knowledge) and a mental level (experience). On both levels, a synthesis is sought, involving the underlying dynamics and specific types of orientation toward reality.

7.9 A Catastrophe Theoretical Model of the Interpretant (Peirce) that Integrates Cassirer's Analysis

Different laws operate on the epistemic and experiential levels. The dynamics of this synthesis define the interpretant. The orientation toward reality can establish a stable link between both types of dynamics.[23] If this advanced notion of semiotic

of an object or object category). In Peirce, the duality introduced by Saussure is the degenerate case of a full-fletched interpretant.

[21] If Peirce seeks for a synthesis of the epistemology of Kant and Hegel in his list of categories (around 1870), Cassirer builds on a critique of Hegel and a new interpretation of Kant on the background of the advances in the natural sciences around 1900 (cf. neo-Kantianism and the Marburg School). He rejects "mental representations" and "representationalism" (cf. Cassirer [23]: 156ff).

[22] Cassirer objects to Hegel's "absolute idealism", that it presupposes a given synthesis between individual-psychological and historical-collective dynamics. By this identification, Hegel's "Panlogicism" drifts toward mysticism (cf. Cassirer [4]: 374). If Berkeley needed the notion of God to bridge the abyss of knowledge and experience, Hegel replaced God with the absolute mind ("absoluter Geist"), the key notion of his phenomenology. Both Kant and Hegel are aware of the opposition between Being and Becoming. However, they did not have the conceptual tools to catch the kinematics and dynamics of Becoming beyond the laws of physics.

[23] The "orientation toward reality" mentioned by Cassirer mirrors the earlier concept of "intentionality" of Brentano and is still a psychological (mental) concept. However, given the threefold ontology introduced in Sect. 7.3, we assume that the "epistemic" level goes beyond any psychological concept. It subsumes (1) the evolutionary process in which humankind was able to get hold

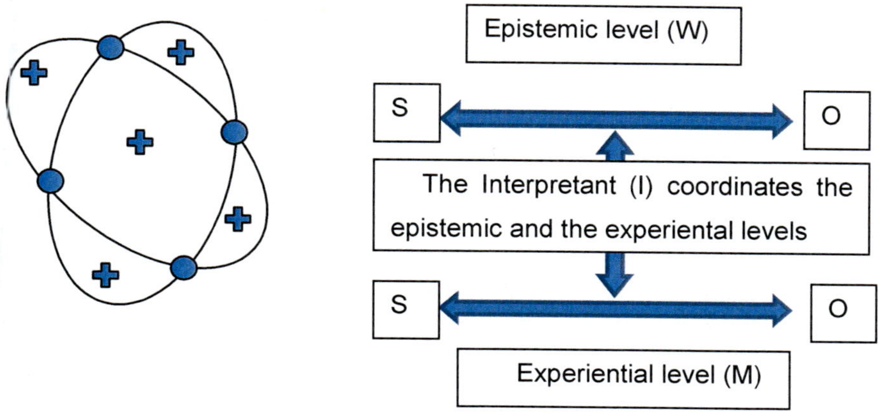

Fig. 7.4 The constellation in the double cusp as a topological schema of the interpretant (epistemic and mental)

synthesis is labeled (with Peirce) "interpretant", then we must give another shape to Peirce's semiotic triangle. In catastrophe theoretic semantics (cf. Wildgen [27]: 85–94 and [28]: 218–222), the constellation of four force fields and the fifth position of synthesis points to the catastrophe called "double cusp". It has four attractors (minima = +), four saddle points (transit-points = o), and in the center, either a maximum (−) or another minimum (+), cf. Callahan [2].[24] This analysis allows for a catastrophe theoretical model that elaborates and completes Peirce's concept of Interpretant (Fig. 7.4).

The transition points between the four attractors (minima, +) mediate between the four force fields (attractors) and constitute a circular flow in which, on one side, dynamic aspects and mental aspects are harmonized. Conversely, the oppositions between sign form and sign object are managed. The Interpretant (I) is at the center of this cyclic process. The details of such a synthesis must be left for further research. This model does not solve the enigma of harmony between the world and mind or

of the permanent aspects of reality and (2) the achievement in prehistoric and historical societies to organize the higher apperception of reality in specific symbolic forms. The diversity of cultures and languages is the outcome of this process. (3) The individual mind can use this heritage to perceive and understand reality. Nevertheless, the success of this understanding remains incomplete and dependent on contexts. Whether future generations will succeed in sharpening or deepening human access to reality or are in danger of regressing from the level already obtained remains open. Nevertheless, even the stabilization of the achieved level of realism can be a success.

[24] In Abraham et al. [1], a complex system of opinion formation in two states in a situation of arms race is modeled. In both states, hawks and doves establish a political opposition. This opposition can be modeled in elementary catastrophe theory with the help of the cusp (a bipolar catastrophe). Suppose both systems are combined and depend on each other. In that case, a double cusp situation is reached, containing up to five attractors, showing different types of bifurcations and limits cycles and chaotic dynamics under specific conditions. However, our model can only be suggestive in this context because neither the epistemic nor the mental dynamics are sufficiently well understood in the current state of the science.

between the epistemic and the experiential view. Still, it shows that, at least at the schematic level, Peirce's triad is insufficient. A model for the dynamics relating the mind, collective knowledge, and an orientation toward reality is due.

The force field between the epistemic level referring to the world (W) and the experiential level referring to the mind (M) and consequently between the double dyad of the sign relation (S, O) shows four types of dependencies:

1. The mind (M) is part of the world (W).
2. The mind (M) is the product of evolution in (W) and adapted to the causalities in (W) (e.g., survival and procreation in the evolutionary theory).
3. Sign communication is usually between different (two or more) minds. However, inner monologue or thought may reduce or replace this social process.
4. Humans use their minds, cognitive abilities, and capacity for symbolic forms to act in the world (W).

The polarity of sign form (S) and sign object (O) is not rigid; instead, it is in motion and underlies continuous changes. Every single use of the sign rehearses and modifies the relational field. As a result, the connection may be reinforced or weakened, the shape of the sign object (S) (e.g., its phonetic or visual shape) can change, and the mental image (I) or the reference (O) can vary.

7.10 Conclusions

Semiotics in Kepler's style contributes to a type of practical realism mostly accepted by working scientists in the natural sciences. In the second half of the twentieth century, Thom, Zeeman, Haken, Prigogine, and their followers proposed and elaborated qualitative models that bridge the bias between the humanities and the natural sciences. Many models in the tradition of linguistics and semiotics labeled metaphorical can be reformulated and reach a scientific status. For instance, the term "valence" is used metaphorically in linguistics and psychology but points to the original use of this term in chemistry that again refers to deep mathematical insights, e.g., Descartes' and Euler's polyhedron formula, Pascal's triangle formula, Gibbs phase rule (cf. Slodowy [24], Wildgen [29]: 45–56 and [35]). Central hypotheses in semiotics and linguistics can be reformulated in mathematical terms, tested for consistency, and prepared for empirical evaluation.

It may seem to be a drawback that the richness of intuitions and theoretical imaginations, i.e., the poetic style in semiotics, would be lost in the Kepler style of semiotics. However, the accumulation of "rich" readings in the poetic style is worthless if no agreement is attainable in the community due to the vagueness and subjectivity of such contributions. Science is a collective enterprise, and all scientific proposals should allow for straightforward critique, response to critique in the sense of clarification and modification, and finally, in the choice of those hypotheses that promise further insights and advancement.

References

1. Abraham, R., Keith, A., Koebbe, M., Mayer-Kress, G.: Computational unfolding of double-cusp models of opinion formation. Int. J. Bifurcat. Chaos **1** (2): 417–430(1991). Internet: https://www.researchgate.net/publication/258992753_Computational_Unfolding_of_Double-Cusp_Models_of_Opinion_Formation
2. Callahan, J.: The double cusp has five minima. Math. Proc. Camb. Phil. Soc. **84**, 537–538 (1978)
3. Carnap, R., Bar-Hillel, Y.: An outline of a theory of semantic information (1952). Reprinted in Bar-Hillel, Y.: Language and information. Selected Essays on their theory and Application. Jerusalem, 221–274 (1952/1964)
4. Cassirer, E.: Das Erkenntnisproblem in der Philosophie der Wissenschaft der neueren Zeit (vol. III). Wissenschaftliche Buchgesellschaft, Darmstadt [first print 1920] (1994)
5. Cassirer, E.: Philosophie der Symbolischen Formen, 3 vol. Wissenschaftliche Buchgesellschaft, Darmstadt (1995) [1923/1925/1929] [English new translation: The Philosophy of Symbolic Forms. Routledge, London (2020)]
6. Chomsky, N.: Syntactic Structures. Mouton, The Hague (1957)
7. Durkheim, É.: Suicide: A Study in Sociology (French original publication 1897). English translation. The Free Press, New York (1951)
8. Eigen, M., Schuster, P.: The Hypercycle: A Principle of Natural Self-Organization. Springer, Berlin (1979)
9. Gibson, J.J.: The Ecological Approach to Visual Perception. Houghton Mifflin, Boston (1979)
10. Haken, H.: Synergetics. An Introduction (third revised and augmented edition). Springer, Berlin (1983)
11. Heidegger, M.: Sein und Zeit [first edition 1927), 11th edn. Niemeyer, Tübingen (1967)
12. Husserl, E.: Ideen zu einer reinen Phänomenologie und phänomenologischen Philosophie. In: Husserliana : gesammelte Werke/Edmund Husserl. Nijhoff, Den Haag. New print 1952 (1913)
13. Husserl, E.: Logische Untersuchungen. Unrevised reprint of the 2nd revised edition 1913 (2011) https://doi.org/10.1515/9783110916089
14. Kepler, J.: Epitome Astronomiae Copernicanae (Gesammelte Werke/Johannes Kepler. vol. 7). Beck, München. First edition 1618–1621 (1953)
15. Kelso, J.A.S.: Dynamic patterns. In: The Self-organization of Brain and Behavior MIT Press, Cambridge (Mass.) (1995)
16. Li, Z., Ji, J., Zhang, Y.: From Kepler to Newton: Explainable AI for Science (2023) https://arxiv.org/pdf/2111.12210.pdf
17. Kleiner, J.: Mathematical models of consciousness. Entropy **22**(6), 609 (2020). arXiv:1907.03223
18. Li, X., Liao, S.: More than six hundred new families of Newtonian periodic planar collisionless three-body orbits. Phys. Mech. Astron. **60**(12), 129511 (2017)
19. Montgomery, R.: The Three-body problem: although mathematicians know they can never fully "solve" this centuries-old quandary, tackling smaller pieces of it has yielded some intriguing discoveries. Sci. Am. (2019). https://www.scientificamerican.com/article/the-three-body-problem/
20. Orwell, G.: Nineteen Eighty-Four 2000. Harvill Secker, London (first published in GB 1949) (1987)
21. Peirce, Ch.S.: Kepler. In: Wiener, Ph.P. (ed.) Charles Sanders Peirce. Selected Writings, chapter 14, pp. 250–256. Dover Publications, New York (1958)
22. Peirce, Ch.S.: Collected Papers of Charles Sanders Peirce, 3rd edn. The Belknap Press, Cambridge (Mass) (1967)
23. Plümacher, M.: Rethinking representation. Cassirer's philosophy of human perceiving, thinking, and understanding. In: Truwant, S. (ed.) Interpreting Cassirer. Critical Essays, pp. 151–169. Cambridge University Press, Cambridge (2021)
24. Slodowy, P.: Platonic solids, kleinian singularities, elementary catastrophes, and lie groups. In: Patitot, J. (ed.) Logos et théorie des catastrophes, pp. 73–98. Patino, Geneva (1988)

25. Thom, R.: Esquisse d'une sémiophysique: Physique aristotélienne et théorie des catastrophes. Interéditions, Paris [English translation: Semiophysics. A Sketch. Addison-Wesley, Boston 1990] (1988)
26. Wildgen, W.: Differentielle Linguistik, Entwurf eines Modells zur Beschreibung und Messung semantischer und pragmatischer Variation. Niemeyer, Tübingen. [Differential Linguistics. A Model for the Description and Measurement of Semantic and Pragmatic Variation] (1977)
27. Wildgen, W.: Catastrophe Theoretic Semantics: An Elaboration and Application of René Thom's Theory. Benjamins, Amsterdam (1982)
28. Wildgen, W.: Archetypensemantik. Grundlagen einer dynamischen Semantik auf der Basis der Katastrophentheorie. Narr, Tübingen (1985)
29. Wildgen, W.: Process, Image, and Meaning: A Realistic Model of the Meanings of Sentences and Narrative Text. Benjamins, Amsterdam (1994)
30. Wildgen, W.: Das kosmische Gedächtnis. Kosmologie, Semiotik und Gedächtnistheorie im Werk von Giordano Bruno. Lang, Frankfurt/Main (1998)
31. Wildgen, W.: Semiotic hypercycles driving the evolution of language. Axiomathes **18**(1), 91–116 (2008)
32. Wildgen, W.: Thom's theory of "saillance" and "prégnance" and modern evolutionary linguistics. In: Wildgen and Brandt (79–100) (2010)
33. Wildgen, W.: Giordano Bruno. Neun Studien und Dialoge zu einem extremen Denker. LIT-Verlag, Berlin [Giordano Bruno. Nine Studies and Dialogues on behalf of an Extreme Thinker] (2011)
34. Wildgen, W.: Cosmological memory space and the origin of human language. In: Hess-Lüttich, E.W.B. (ed.) Sign Culture-Zeichen Kultur, pp. 1153–1167. Königshausen & Neumann, Würzburg (2012)
35. Wildgen, W.: Structures, archetypes, and symbolic forms. In: Peruzzi, A., Zipoli Caiani, S. (eds.) Structures Mères. Semantics, Mathematics and Cognitive Sciences. Applied Mathematics in Linguistics and Semiotics, pp. 165–185. Springer, Cham (2020)
36. Wildgen, W.: Mythos und Religion. Semiotik der Transzendenz. Königshausen & Neumann, Würzburg [Myth and Religion. The Semiotics of Transcendence] (2021)

Chapter 8
Final Reflections

Abstract This final chapter reflects possible criticisms of the assumptions in this book and tries to evaluate their weight and the plausibility of the strategies chosen. The strategical doubt concerns, first, the application of the concept of morphogenesis to linguistics and semiotics (theory of symbolic forms), second, the worth of mathematical models, and more specifically, the preference for topological and dynamic models sustained in the central Chaps. 3, 4, 5, 6 and 7, third, the relevance of Cassirer's hypothesis that the different symbolic forms present parallel modes of understanding the human world, and fourth, the fecundity and practicability of the Kepler methodology and the strategy of naturalization in the humanities.

Morphogenesis is defined and exemplified mainly in biology, such as embryology and gastrulation. It describes the first but decisive form giving developments of a complex, mainly a vertebrate body. Secondly, in applied differential topology (catastrophe theory), the term morphogenesis is linked to the "unfolding" of elementary dynamical systems (differential equations) to structurally stable bifurcation diagrams. This mathematical level of abstraction builds a bridge between simple biological systems and cultural phenomena, for instance, the symbolic forms constitutive of human communication and culture. The epistemological itinerary is abductive (in the sense of Peirce), i.e., on the one side, empirical facts in biology and the humanities or cultural sciences provide a corpus of descriptions and some trends toward explanation. On the other side, the formal results of mathematics give a coherent and well-organized architecture of concepts that allows for deductive reasoning. If both strategies, empirical induction and mathematical deduction, are combined correctly, a new scientific understanding becomes accessible.

Our enterprise, "morphogenesis of symbolic forms," is based on three assumptions. The course of the argument and the examples analyzed in this book tried to validate these premises. Nevertheless, they were not proven or tested experimentally or based on a vast corpus of data. We shall, therefore, test their stability in confrontation with doubt (playing the role of an "advocatus diaboli"):

W. Wildgen, *Morphogenesis of Symbolic Forms: Meaning in Music, Art, Religion, and Language*, Lecture Notes in Morphogenesis,
https://doi.org/10.1007/978-3-031-25651-6_8

(a) *Morphogenesis.* Can this concept (with its applications) be transferred from ovulation, gastrulation, and similar biological standard examples to the evolution of species, mainly the human evolution, the unfolding of the cognitive and social capacities of humans, and finally, the symbolic forms that organize human communication and culture? Are the formal, geometrical, and dynamic principles underlying biological morphogenesis (cf. Thom's work) also applicable to cognitive (eg. human) evolution and cultural, symbolic evolution? In the biological developments and geology field, such evolutions are visually accessible. At first view, this seems impossible in the case of cultures and symbolic forms. Here, a secondary premise comes into play. Morphogenesis relies heavily on spatial and temporal factors. If symbolic forms refer to space and time, then the assumption of visual accessibility gains some plausibility. But is this assumption corroborated by empirical studies? The examples elaborated in Chaps. 2, 3, 4, 5 and 6 provide some evidence but cannot cover all areas. Therefore, this premise remains plausible but unproven.

(b) *Relevance of mathematical, specifically geometrical/topological models.* This argument has received support since the early nineteenth century (in psychology) and later in sociology and linguistics (mainly in the twentieth century). The trend became very strong with the use of computers after the Second World War.[1] This rich tradition shows that without such endeavors, modern science would stagnate. In the case of dynamic systems theory, one can argue that the broadness of successful applications in the natural sciences, including biology and cognitive sciences, hints at these formal devices' potential. They are less abstract than logics, Bourbaki structuralism (cf. [1]), or automata theoretical linguistics (Chomsky) and rely on a rich tradition of successful applications since the introduction of differential calculus by Newton and Leibniz at the end of the seventeenth century.

(c) The *philosophy of symbolic forms.* Is there a common denominator for language, myth, science (cf. the titles of Cassirer's trilogy 1975, published 1923–1929; [4]) and further manifestations, such as art, technics, ethics, politics, and economy (cf. for an analysis of these fields Sandkühler et al. [6]? Our hypothesis (in Chap. 2 was that the sensual apparatus (the human body and its contact zone with human ecology) is the baseline from which the morphogenesis of symbolic forms starts. The differences in these windows towards reality are the causes for the differences between the symbolic forms. The auditory sense is the window to music, and the visual sense is the window to art. Both senses give access to the use of language in its spoken and written manifestations. The myth may take its "material" from different senses; it further activates a kind of inner sense

[1] Unification programs in mathematics and logic and programs of a unified science appeared at the beginning of the twentieth century. This trend is visible in Peirce, who proposed to use relational logic as a fundament of semiotics; the mathematical logics of Whitehead and Russel [8] were the basis of the proposals of Carnap in his "Formal Syntax" [2] and later of Chomsky's generative grammars. Bourbaki's meta-mathematics based on set theory was used by Jean Piaget in his "Genetic Psychology" [5] and the mathematical unification program of "category theory" was applied in diverse linguistic and semiotic enterprises' (cf. Wildgen [7]).

expressed by emotions. A classical critique of such a mild "sensualism" is that perception's functional and social embedding could be responsible for the differences between symbolic forms. It may even have molded animal and human perception and cognition capacities. This argument is an aspect of the classical nature-nurture debate. The assumption of morphogenesis and, for complex organisms, the assumption of self-organization does not exclude contextual, functional effects. It instead assumes that there is a temporal and processual priority for the body- and world-rooted determinations. The first steps of evolution follow general morphogenetic principles; later, the accumulated complexity requires self-organization to achieve stability and permanence. Contextual and functional aspects are relevant throughout, but their effect becomes dominant later when a stable fundament of animal or human bodily organization is already given. The system can "pay" for complicated, even aleatory (noisy) additions. In crisis episodes, this richness will disappear as quickly as it has been added. Therefore, the scientific strategy should aim to discover the stable and essential aspects of human cognitive and cultural capacities.[2]

(d) The *naturalization* in the humanities or "Kepler strategy". Kepler, like most working scientists, assumes a naïve realism; he tentatively trusts his senses, although he knows their limits (of confidence). He works hard to enlarge the field of confidence by using and devising better instruments and with the help of better exploitation of large data sets (thus in his analysis of Mars orbit). In the case of astronomy, an ontology of the visible (1), i.e., a realism based on the intelligent use of the available techniques of observation and experimentation, may be sufficient. Semiotics, or the analysis of symbolic forms, ask for a broader basis, including at least the cognitive and emotional basis of the (individual) human mind (2) and the integrative effects of social perception and action (3). The ontologies (2) and (3) depend on each other and the ontology of objects in the natural sciences (1). As the debate on (global and local) ontologies has remained open in philosophy, one must concede that our proposal of three levels (in the case of symbolic forms) is only a methodological hypothesis enabling a more transparent treatment than philosophical or metaphysical presuppositions. Husserl's notion of "local ontologies" comes near to our purpose, i.e., in the context of our analysis of the morphogenesis of symbolic forms, this position is plausible and practical. It negates a physicalist position (physical monism), the Cartesian dualism (mind-matter), and the chaos of a myriad of subjective ontologies. Our ontological levels span the field between physics and the cultural sciences; the cognitive sciences *animate* the zone between both. Chapters 3, 4, 5 and 6 accumulate evidence that a strictly non-monistic and

[2] Cassirer discusses the debate on Darwinism in the fourth and last volume of "Das Erkenntnisproblem in der Philosophie und Wissenschaft der Neueren Zeit" ([3]; first edition of this volume in English 1950 under the title "The Problem of Knowledge. Philosophy, Science, and History since Hegel"). Unfortunately, his review stopped in 1932; the new synthesis of Darwinism due to advances in statistical modeling and modern genetics was not considered. Therefore, Cassirer is not a reliable source of evolution questions and underestimates the nearness of human and animal cognition and communication.